Health Ecology

The world is a highly interactive, dynamic and adaptive system. The impacts of human existence and activities on this system and on the health of humans have become a major concern of our time. Yet conventional discipline-based investigations do not explain the complexities of phenomena such as health, culture and human–environment interactions. The concept of Health Ecology responds to such conditions through its holistic approach, enabling a clearer understanding of the roots and dimensions of the issues concerning the advancement of health and the creation of a healthy environment.

Health Ecology presents an introduction to examining health from a human ecological perspective. Bringing together a variety of approaches from different perspectives and different locations, the contributors examine the various dimensions of health ecology in a human ecology framework, looking at how local, regional and global factors impinge upon the health and environment of individuals, communities and the globe.

Experts from human ecology, public health and policy, sociology, anthropology and geography draw on a range of global examples – spanning Australia, Indonesia, Spain, Belgium and North America – to examine issues such as health in urban and rural areas, child health from an ecological perspective, healthy homes, health within a political context, health and sustainable development, and health promotion and the media.

Setting out new foundations for Health Ecology, which aim to create and maintain a sustainable state of health for human beings in a healthy environment, this book offers new challenges to those teaching, studying or developing strategies and policies in health and the environment.

Morteza Honari has recently established a consultancy and advisory service centre on health, environment and culture in Sydney, Australia. **Thomas Boleyn** is the convenor of the postgraduate course in Primary Health Care at the Faculty of Health Sciences, University of Newcastle, Australia.

Health Ecology

Health, culture and human–environment interaction

Edited by Morteza Honari and Thomas Boleyn

London and New York

First published 1999
by Routledge
2 Park Square, Milton Park, Abingdon, Oxon, OX14 4RN

Simultaneously published in the USA and Canada
by Routledge
270 Madison Ave, New York NY 10016

Transferred to Digital Printing 2005

Typeset in Galliard by M Rules

British Library Cataloguing in Publication Data
A catalogue record for this book is available from the British Library

Library of Congress Cataloging in Publication Data
Health Ecology : Health, culture, and human–environment
  interaction / edited by Morteza Honari and Thomas Boleyn.
    p.   cm.
  Includes bibliographical references and index.
  1. Environmental health.   2. Human ecology—Health aspects.
  I. Honari, Morteza.   II. Boleyn, Thomas.
  RA565.H39   1999
  306.4'61—dc21   98-30278 CIP

ISBN 0-415-15446-4 (hbk)
ISBN 0-415-15447-2 (pbk)

Printed and bound by Antony Rowe Ltd, Eastbourne

Contents

List of figures   x
List of tables   xii
List of contributors   xiii
Foreword   xv
ZENA DAYSH

1   **Health ecology: an introduction**   1
MORTEZA HONARI

Introduction   1
Ecology   2
Human ecology   4
Health   19
Health ecology   22
Health impact assessment   23
Health ecology   24

PART I
Health in macro ecosystems   35

2   **Good planets are hard to find**   37
ILONA KICKBUSCH

The concept of an ecological public health   37
Societal developments   38
The societal response to health risk patterns   39
A new public health agenda   41
From the sewerage principle to an ecological principle   43
Theoretical base of an ecological public health   45
The pattern that connects   47
Strategies and approaches   48

Consequences of a new public health strategy 50
Changing expectations and social perceptions 51
Political ecology 52

3 Health and conservation: shared values 59
DAN W. WALTON AND PETER BRIDGEWATER

Introduction 60
Principal components 63
Ecology and the environment 65
People 67
Discussion 69
Conclusion 73

4 Human health as an ecological problem 79
NAPOLEON WOLANSKI

A conceptual overview 79
Environment 82
Organism 85
Organism–environment relation: a criterion of human health 91
Genetic determinants and environment 93
Positive indices of health 94
Negative indices of health and cultural maladjustments 101
Contemporary civilisation-affected diseases 104

5 Health through sustainable development 112
JERZY KOZLOWSKI AND GREG HILL

Defining sustainable development 112
Sustainable development and sustainable health 113
Planning for sustainable health: the potential 120
Planning for sustainable health: the tools 124
The Ultimate Environmental Threshold method 127
Conclusions 130

6 Health and political ecology: public opinion, political ideology, political parties, policies and the press 135
NANCY MILIO

Introduction 135
The climate of policy making in the 1990s 136
Public opinion 136
Political ideology 138

The press	142
Political parties and other stakeholders	146
Policy impacts on health	147
Conclusion	148

PART II
Health in micro ecosystems **151**

7   **Health of women: changing lifestyles and reproductive health** **153**
CRISTINA BERNIS

Introduction	153
Health and reproductive health	155
Secular changes in the biological determinants of fertile life	157
Behavioural determinants of fertile life use	161
Fertility changes over time in Spain	164
Socio-economic variation	165
Methods of birth control: type, medical surveillance and failure	166
Health consequences of change in reproductive patterns	169

8   **Health of children: causal pathways from macro to micro environment** **175**
NICK SPENCER

Introduction	175
Brief overview of micro environmental explanations	176
Brief overview of macro explanations	178
Mediators linking the macro to the micro	179
Temporal, inter-generational and cumulative effects	181
Causal pathways from macro to micro environment	182
A theoretical framework for the analysis of causal relationships in child health	187
Summary and conclusions	189

9   **Healthy homes** **193**
HOSSEIN ADIBI

Introduction	193
Characteristics of healthy homes	194
Conclusion	204

PART III

Selected case studies   207

10 Health ecology and the biodiversity of natural medicine:
    perspectives from traditional and complementary
    health systems   209
    GERARD BODEKER

    Traditional health systems: policy, biodiversity and global
        inter-dependence   209
    The global context   210
    Health care costs and traditional health services   214
    Research policy in traditional health care   215
    Biodiversity   218
    Conclusions   223

11 Health of rural and urban communities in developing
    countries: a case study in Indonesia   227
    SHOSUKE SUZUKI

    Introduction   227
    An overview of the issues in developing countries   228
    Population of developing countries and their environment   230
    Health and health outcomes in rural Javanese villages   232
    Health and health outcomes in urban Jakarta: air
        pollution   243
    Conclusion   248

12 Health and psychology of water   250
    DAVID RUSSELL

    Introduction   250
    Water as a scientific and cultural construct   250
    The historical duality of water   251
    Waste water and reclaimed water   252
    Reclaiming our dreaming   254
    Designing a social ecology of water   254
    Conclusion   256

13 **Health Impact Assessment in Flanders: contribution to environmental management and health** 258
PETER JANSSENS AND LUC HENS

Introduction 258
Guidelines for Environmental Health Impact Assessment in Flanders 260
Elements of the process of EHIA 260
Post-monitoring and post-evaluation 265
Conclusion 265

Index 267

Figures

4.1 Zones of biological tolerance of species, in relation to climatic zones   84

4.2 Model of eco-sensitivity and adaptability of organisms to environmental stimuli   86

4.3 Model of adaptational changes of an organism in response to environmental stress (bio-adjustment)   88

4.4 Adaptive changes of cardio-respiratory functions of humans, measured in various climatic conditions in industrial and urban settings in Poland   89

4.5 Model of adaptive behaviour in modernisation: natural and socio-cultural environments   90

4.6 Haemoglobin concentration plotted against hematocrite index in inhabitants of specified locations in Poland   98

4.7 Infant mortality per 1,000 live births in Poland, 1946–1995, compared with annual percentage change in GNP   102

4.8 Infant mortality rates in selected European countries, 1960, 1975 and 1985.   103

6.1 Primary policy-making processes affecting life and health of populations   136

7.1 Structural components of human ecosystems   154

7.2 Biological and social constraints on fertility   156

7.3 Prevalence of early maturers, experiencing menarche at age 12 or earlier, in selected Moroccan and Spanish populations   158

7.4 Distribution of the interval between the birth of a last child and onset of menopause among women in urban Spain and in rural Morocco   160

7.5 Variability of fertile life in selected Moroccan and Spanish populations   163

7.6 Mean number of menstrual cycles experienced by Moroccan and Spanish women   164

7.7 Variation in age at marriage in Spain, 1900–1985   165

7.8 Variability in some bio-social determinants of fecundity according to the level of education of Spanish women   166

7.9   Age distribution of contraceptive failure in selected Spanish
        populations   168
8.1   Three-tier causal model of child mortality   183
8.2   Path model showing cultural, socioeconomic and medical
        predictors of infant death in 480 Sri Lankan households   184
8.3   Inter- and intra-generational relationships between health and
        social circumstances   185
8.4   Hypothetical path to survival in a neonatal intensive care unit   186
8.5   Final path analysis model   186
8.6   Model of relationship between deprivation, IQ, temperament
        and offending in adolescence   186
8.7   Theoretical causal framework for the determination of child
        health   188
11.1 Life expectancy and GNP per capita for selected countries in
        1991   229
11.2 Map of Indonesia, showing major islands and regions   233
11.3 Rice fields in West Java   234
11.4 Map of Salamungkal, a village in West Java   236
11.5 A washing place by a pond   237
11.6 A typical well used in villages in West Java   238
11.7 A farmer and his daughter-in-law fish, with a net, from a latrine
        pond   241
11.8 Urban population as a percentage of the total population of
        selected countries   244
11.9 Health effect of lead in atmosphere polluted by automobile
        exhaust gas   247

Tables

4.1 Mechanism for maintaining equilibrium between a population or
     an organism and the external environment   87
4.2 Recommended selection of variables for the monitoring of human
     biological status   96
5.1 General classification of human needs   117
7.1 Main differences in reproductive patterns between rural and
     urban ecosystems   162
7.2 Type of contraceptive used in relation to the educational level of
     women   167
7.3 Frequency of contraception use and its failure, by size of
     population   167
10.1 Relationship between traditional and modern medicine   211
10.2 Old and new perspectives on traditional systems of health care 223
11.1 Vicious and healthy cycles of 'education–population–resource
     production'   231
11.2 Population area, and population density by region in Indonesia 234
11.3 Quality of drinking water and washing water used in Salamungkal 240
13.1 Overview of projects subject to Environmental Impact Assessment
     in Flanders, Belgium   259
13.2 Six steps of Environmental Health Impact Assessment   261
13.3 Zaponi list of data for use in evaluating the health impact of
     chemical substances   263

Contributors

Zena Daysh, of the Commonwealth Human Ecology Council, London, has been a pioneer in applied human ecology since the 1950s.

Ilona Kickbusch was the key instigator of WHO's Ottawa Charter for Health Promotion and Healthy Cities project. She is Professor of International Health at Yale University.

Peter Bridgewater is a landscape ecologist and the head of the Australian Nature Conservation Agency, Canberra.

Dan W. Walton is Senior Adviser at the Australian Nature Conservation Agency in Canberra, Australia.

Napoleon Wolanski is Professor and Head of Department of Human Ecology at the Polish Academy of Sciences, Warsaw, Poland, and also at Departamento de Ecologia Humano, CINVESTAV, Yucatan.

Jerzy Kozlowski is Professor of Town and Regional Planning at the Queensland University, Brisbane, Australia.

Greg Hill is Professor of Tropical Environmental Science at the Northern Territory University, Australia.

Nancy Milio is Professor of Health Policy and Administration at the University of North Carolina, Chapel Hill. She has devoted her life to the enhancement of health policy, health promotion and health education.

Cristina Bernis is Professor of Biological Anthropology at the Universidad Autonoma de Madrid, Spain, and Editor of the *Journal of Human Ecology*.

Nick Spencer is the Foundation Chair in Community Health and Head of the School of Postgraduate Medical Education at the University of Warwick, UK.

Hossein Adibi is a Sociologist based in the Department of Social Sciences at the Queensland University of Technology, Brisbane, Australia.

Gerard Bodeker is Chairman of Global Initiative for Traditional Systems of Health, University of Oxford.

Shosuke Suzuki is Professor of Public Health at Gunma University in Showa, Japan.

David Russell is Associate Professor of Social Ecology at the University of Western Sydney, Australia.

Peter Janssens is a biologist based at the Free University of Brussels, Belgium.

Luc Hens is Professor of Human Ecology, Free University of Brussels, Belgium.

Thomas Boleyn has devoted his life to education for and practice of Primary Health Care. He is an advocate of a humane approach to health care, and the co-editor of this book.

Morteza Honari, a human ecologist, has spent his life in developing a holistic approach to culture, health and nature. He has initiated the concept of health ecology, and has edited and coordinated the development of this book.

Foreword

Human ecology and health in a global system

Zena Daysh

The editors have requested that I contribute a foreword to this collection. This text also encapsulates a summary of the history and work of the Commonwealth Human Ecology Council (CHEC), the organisation of which I was a founder member forty years ago, and of which I am now head.

Human ecology

Human ecology, assessed in all its meanings, highlights health, which at all times provides a penetrative and influential thread.

It is worth stressing that human ecology is an amalgam of disciplines with a canon and corpus of knowledge. Yet it is still a unity which provides a logical and powerful framework of human action in its own right.

Human ecology is the totality of concerns relating to the human-made and the environment and their interactions. This embraces the holistic view of individual responsibility and the inter-relations of people and their surroundings. Human ecology as a field of study of human life steps beyond the confines of a narrow discipline to be able to see the whole. This is never easy, especially as traditional education emphasises scientific disciplines and a narrow subject-based learning.

The pioneers of human ecology, forward thinkers and leaders in their fields, came together after the Second World War, in London, the capital of a fading empire. They recognised a well-advanced global system emerging after 300 years of imperial history, maturing into a modern Commonwealth where they could focus their cooperation on the study and practice of human ecology, since it was already understood that human ecology as a global concept required globally oriented practical action.

The Commonwealth, with one-third of the world's population and a quarter of its area, provided the first practitioners of human ecology with a large, even overwhelming, field of study, a wide vista of perspectives and the widest variety of traditions and cultures, with institutions which reflected them. By focusing on the connecting threads of the human being, society and the global environment, this comprehensive grouping allowed for a new definition of human ecology and health from a more universal perspective.

Human ecologists often describe their approach by saying '*everything connects with everything*'. This idea found common agreement among the participants at the International Human Ecology Conference held in Tokyo, organised in 1988 by the Department of Human Ecology in the Medical Faculty of the University of Tokyo, where that phrase became our motto.

I recall being deeply satisfied with this consciousness, since the *raison d'être* and the context of the work of the CHEC was developed on just such an awareness.

CHEC also recognised that development in a nation is very slow process under normal conditions, and it takes at least two generations to achieve a fundamental change in values and in the resulting emotional changes. *Sustainability*, which has now become one of the most important concerns, reveals the power of the ecological approach in the long term and the value of its concepts.

Now more than ever, we have come to realise that our large, and at times even overwhelming, field of study – human ecology – especially when seen in a practical, global and political perspective – provides a long enough vista of perspectives, in a systemic context, to connect all the threads of human being, society and the global environment, presenting *health ecology* as a new challenge.

This book sets out to lay a foundation for health ecology, which aims at creating and maintaining a sustainable state of health for human beings in a healthy environment.

Components of health

Health has always informed my approach to development, but I have resisted the narrow view of health as a technical or medical matter. I see health in its broadest sense as a matter of energy, upheld by a spiritual force which necessarily encompasses all aspects of the life of individuals and communities, and their environments.

The nature and acceptance of inner growth requires us to examine our attitudes of mind. Through this we can learn humility and patience which help us to overcome stress and other unhealthy consequences. Self-awareness also forces us to recognise the importance of being humanitarian, since we are responsible for many of the situations we find ourselves in. By avoiding selfish living we ensure a safer environment and better social surroundings which rebound on us personally.

Health control is a matter of deeper consciousness. *Disease* is what it says, namely, *absence of ease* and an imbalance of the self. Maintaining health also requires balance, the base of which is discipline. There is a price to pay for excess in behaviour and activities which destroy balance and cause malaise. There is a need in sustainable development for discipline to build a machinery of skills and institutions which are required to maintain health and the health of all.

Health is a global affair – who would think of a state of health separated from the holistic sense that *everything connects with everything*?

Health in CHEC programmes

CHEC is probably the first explicitly human ecological non-governmental organisation in the world, and certainly the only such NGO to have consultative status with the UN Economic and Social Council, ECOSOC (1972–1998).

CHEC started its life in the 1950s in the London School of Hygiene and Tropical Medicine, a leading postgraduate university of international repute. There, generations of public health, social and medical scientists with community health experience, coming from all corners of the world, worked together to widen their knowledge, while contributing to the unique understanding of health and well-being for many cultures and climates. This laid the bases for health care policies in many of the countries which make up today's Commonwealth. Such balanced development, improving community health, is facilitated globally through the School's close partnership with the UN Centre for Human Settlements (Habitat).

This group of academics from 37 disciplines came together, after an uphill struggle of two or three years of intellectual and field developmental research, to form a research group centred on the unifying concept of nutrition. The notion of '*holistic health*' guided us to establish the Committee on Nutrition in the Commonwealth. The innovative nature of our work revealed health as an amalgam of balanced nutrition, a disciplined mind, a balance of energies, a humanitarian approach and individual responsibility, participating in an improving socio-economic framework.

The Committee on Nutrition in the Commonwealth attracted experienced social and preventive medical scientists and academics from the new nutrition departments of the universities; later, agronomists, biologists and experts from other natural sciences came on board. We were then joined by geographers, political and social scientists, including anthropologists, and finally embraced theologians and philosophical advisers with wider perspectives.

Our cooperation showed that the single disciplinary approach could not stand up against the multi-disciplinary approach to human health, to develop programmes improving the balance of human and societal well-being.

In 1970 the Committee on Nutrition in the Commonwealth became the Commonwealth Human Ecology Council (CHEC). Following lengthy and comprehensive research based in human ecology, CHEC gained a sound international foundation – benefiting from the unique historical, emotional, political, cultural and economic ties linking Commonwealth countries, and making them receptive to the benefits of wider perspectives.

Our first international project in Malta kept health to the fore. This small, newly independent Mediterranean country hosted a project involving a range of government departments (Health, Agriculture and Education) and academics including social scientists, architects, planners, lawyers, natural scientists, medical researchers, and health researchers. Our findings provided the indicators we sought and comprehensive health measurements of the population.

In the years to follow, fifteen more national programmes came into existence, often in the form of and administered by national CHEC chapters, which now include: Australia, Bangladesh, Barbados, Canada, Guyana, India, Indonesia (Bali), Kenya, Malawi, New Zealand, Pakistan, Sierra Leone, South Africa and Sri Lanka. Major community programmes have been put into practice dealing with the basics of individual and societal development.

Health ecology

In this book, the concept of health has been examined and shown in operation from many directions. The editors and their colleagues have provided theoretical and practical foundations for the subject. This book provides a new beginning in examining health from a human ecological perspective, a new challenge for all who teach, study, research, and develop strategies or action plans in health and well-being and in the best spirit of human beings and their environment.

1 Health ecology: an introduction

Morteza Honari

Abstract

The term *health ecology* is coined for and creates a focus for the study of fundamental, complex and inter-related factors. Such factors, whether of a global or individual nature, impinge upon the health and environment of individuals, communities and the globe.

This chapter examines concepts, principles and practices of sustainability and analyses the range of indicators which provide the ground for a holistic approach towards the advancement of health and the creation of a healthy environment.

This introduction links together theories, concepts, applied projects and case studies in order to outline major dimensions of health ecology.

It highlights the conceptual pathway from ecology to health ecology, and links this to my own personal journey.

Introduction

A holistic approach has become the requirement of our time which enables us to find the roots and dimensions of the issues of human health, culture and environment. There is a growing understanding and acceptance at the personal and the intellectual level that everything is linked to everything else.

We have come to realise that the world is actually a highly interactive, dynamic, non-linear adaptive system. It has become increasingly clear that traditional studies do not carry within themselves adequate paradigms for an understanding of complex phenomena such as health, culture and human–environment interactions. It is most appropriate that the health component of this complex system be an object of integrated study from a human ecology framework.

Health and *environment* are transdisciplinary, multi-dimensional and holistic in nature. It is their inter-relationship, interdependence and interplay that are the issues of intellectual, professional and public concern. These concerns have brought about the necessity for new thinking, vision and action. New approaches are required to soften boundaries, to integrate disciplines, to comprehend complex issues, and to develop strategies for non-linear problems.

The development of the concept of *health ecology* is a reflection of such needs at a theoretical and practical level.

For me, the search for a holistic approach is a lifelong pilgrimage. I was born into a family of learning and teaching. I have become of necessity a student of geography, economics, geology, history, anthropology, environment, literature, folklore, religion, philosophy, poetry and theatre. I have learnt as I continue to struggle to learn from these disciplines. I have had some great teachers. I have been lucky in this sense.

Discipline-based approaches have never satisfied me. I have always felt that something was missing; that there were issues of great significance being omitted. Any invitation to examine and acknowledge such an approach generates narrow-mindedness, ignorance, lack of self-knowing, absence of wisdom, the parading of limiting stupidity as arrogance. In the mid-1970s, when I first came to know about human ecology and its holistic vision, I felt at ease.

The human ecology perspective is holistic because of its nature, literature, methodologies, and ultimate goals. As Gerald Young suggests, 'it is the ecology of ourselves' (1991). Human ecology adapts its sources of information and methods to suit the dynamically changing patterns of nature and culture, within and surrounding human beings. Human ecology has the potential to place health at the centre of human and environmental interactions. This to me is health ecology.

This term, health ecology,[1] is a result of my lifelong struggle to find a concept and framework with the capacity to encompass many disciplines that are themselves complex and multi-dimensional.

Although I have provided a review of conceptual developments from ecology to health ecology, it is not my intention to give comprehensive definitions, but to highlight those concepts which provide me with a foundation for health ecology.

Ecology

The study of life, except human life, is located in biology. Ecology was born into the discipline of biology. Hanns Reiter appears to have been the first to combine the Greek words *oikos* (house) and *logos* (study of) to form the term *ecology*. There is consensus, however, that Ernst Haeckel was the first to give substance to the term by defining it as the study of 'the total relations of organisms to the *external world*' in 1866.

Haeckel defined his use of the term ecology as 'the body of knowledge concerning the economy of nature – the investigation of the total relations of animals both to their organic and inorganic environment. Ecology is the study of all the complex inter-relations referred to by Darwin as the condition of *struggle* for existence.'

Since the 1960s, ecology has become a household word with a simple accepted definition: 'it is concerned with the relationship between plants and animals and the environment in which they live'. When the word is used, many people think of it as another term for pollution, standard of living, conservation of endangered species, opposition to development projects or a political plot against economic growth.

With the rise of ecology, some concepts gained momentum: environment, relationship, and holistic approach.

Environment

Within the definition of ecology, there is recognition of a cultural space in which everything interacts with everything else. The concept of environment covers just about everything. Everything which is there; in a pond, a garden, or on the earth. Everything which is associated with an organism at the centre is its ecology. This includes other organisms and the non-living part of the world in which life occurs. The weather, the physical and chemical composition of the soil, and seasonal changes in the length of daylight, are all parts of an organism's environment. Somehow the notion of time is interwoven into it as well. It changes. It is dynamic. It evolves. No organism exists without an environment. Environment, regardless of its size and dimension, is the arena for ecology.

Relationships

Another central concept in ecology is 'relationship between', which implies there is something continuously happening. There is an emphasis upon the concept of dynamism. This notion has since been developed into many terms which express this dynamism, process and chain of events, such as interaction, inter-relation, inter-dependency, interplay and linkage. Everything depending on everything else.

A holistic approach

The greatest challenge facing ecology, even from its early years, is its approach. It brings about a framework for thinking holistically. It is not possible to take a narrow reductionist approach in the study of living organisms and non-living elements, in a given place, at a given time, and to study all interplays between them. It requires holistic thinking. There is something beyond a specific plant, animal, matter or individual. It is the nature of totality. We must be prepared to develop and sustain a vision of totalities.

As the term ecology became accepted as a unified study of the inter-relationships within environments, the dynamism of relationships has been examined at different levels. Living members of any species in one environment constitute one population. The totality of living species in one environment is defined as one community. Ecosystems are either networks of interactions within a community or those inter-linkages with the physical environment; they can be both. The term ecosystem, which was introduced by the British botanist A.G. Tansley in 1935, has been used to express the circulation and transformation of materials through various parts of an environment, thus demonstrating the interactional interdependence of living and non-living elements.

While ecology was evolving in its own way there were attempts to learn from ecology and to apply its frameworks to human beings. We, as humans, struggle to understand life and identify our own niche in it. It has been a continuous and enlarging effort. As human perception of life evolves, the challenge to understand those conditions in which we live continues.

Human ecology

Human ecology emerged from biological ecology through sociology. The term 'human ecology' was first used by sociologists Park and McKenzie in the early 1920s. They were studying the growth of cities, and trying to apply principles of biological ecology to their dynamic transformation and growth.

History

Although human ecology derived its name and framework from the well defined discipline of biological ecology, there are people who prefer to consider the concept rather than the name, and it has been suggested that: 'while the term "human ecology" is new, the concept is as old as human life on the Earth'. Zena Daysh, a standard bearer for human ecology since the early 1950s, suggests that: 'to see human ecology in a comprehensive historical perspective, it pays and enlightens to go back to the ancient roots and experiences of earlier humans in their natural world' (Daysh, 1988). An account of the developmental history of human ecology gives some indication of the process.

The early years

In Chicago in the 1920s, during the boom and busts cycles of the American economy, Robert Park and Roderick McKenzie coined the term 'human ecology' in their paper, published in the book *City*. This paper was based on the theories of sociology of Herbert Spencer and Emile Durkheim. Park and McKenzie were exploring applications of those principles by which plants and animals live in their environments, when applied to the dynamisms of human interactions within their physical urban environs.

Their work was a combination of intellectual, professional, commercial, and integrative ideologies. Human ecology, during this period, was tightly focused on discovering general rules of urban environmental transformation.

Park in 1925 suggested the typical constellations of persons and institutions in space as the main study of human ecology. McKenzie added that 'human ecology should try to study spatial and temporal relations of human beings affected by selective, distributive, and accommodative forces of the environment . . . and to study the effect of position in both space and time upon institutions and human behaviours'. This focus was more on the built environment than on the natural environment as an important factor in human existence.

Park later clarified his view by considering biotic balance and social equilibrium in societies and communities. He suggested that: 'It is the inter-action of four factors that maintains the biotic balance and the social equilibrium.' He named the four factors as population, artefacts (technological culture), customs and beliefs (non-material culture), and natural resources (Park, 1935).

McKenzie also explained his view further by suggesting that:

> The spatial and sustenance relations in which human beings are organised are ever in the process of change in response to the operation of complex environmental and cultural forces. It is the task of the human ecologist to study these processes of change in order to ascertain their principles of operation and the nature of the forces producing them.
>
> (McKenzie, 1926)

There were other developments that contributed to exploring a niche for human beings in nature, sometimes even using terminologies more familiar in sociology.

In 1926, zoologist J.A. Thomson wrote a series of articles for the *Quarterly Review* on the relationship between biology and social life. He described three areas: folk, work, and place. Folk covered life itself, sexual reproduction and the physical and mental quality of the population. Work involved occupational health and industrial efficiency. Place was concerned with the improvement and beautification of the environment (Thomson, 1926).

In 1934 Julian Huxley described biological and social engineering in much the same terms. He believed that the broad political ideas on which society was founded needed to be supplemented by science, in particular the science of heredity and psychology. This implied that, in the future, everyone would be subjected to measurement of physique, temperament, intelligence, constitutional proneness to disease, vocational aptitudes and special gifts (Huxley, 1934).

The 1950s and 1960s

At the end of the Second World War, at a time of restructuring and rebuilding, human ecology spread to European universities through research in population and urban studies. There were also descriptions of the distribution of persons and institutions in space, as well as efforts towards discovering spatial trends of population settlement, and those forces organising communities both in space and time.

Amos Hawley published his book *Human Ecology* in 1950, and in it he defines the subject as the study of the form and development of the community. Success is measured by the community's capacity and capability for population growth, because in such manner the community shows that it is able to solve functional problems through specialisations, internal work divisions, and technological developments.

In the 1960s in Europe, the study of urban areas was strongly conceptualised in terms of neighbourhoods, community activities and behavioural changes in urban environments. Some studies examining industrial relations were also conducted under the label of human ecology.

After the Stockholm Conference

The energy crises of the late 1960s brought about new issues of concern for human ecology. There followed a crisis of resources and a recession in metropolitan growth, as well as economic crises.

Prior to the 1970s, the main concerns of human ecology were the expansion of technology, urban areas, institutions and populations. Human ecology was greatly enlarged during this period as an inter-disciplinary field. This period was concerned with the change in patterns of human relationships with the natural environment and resources; from consumption to conservation and to actions of redemptive stewardship.

Another significant development was the changing focus of human ecology from problems of urban industrial countries to those facing poor and underdeveloped countries. This shifted the emphasis from pollution caused by industries and over-development to poverty and under-development – which are equally destructive to the physical and to the human environment.

Nothing exposes differences of emphasis more clearly than the stated purposes of two organisations, both using the label of human ecology. The Society for Human Ecology (SHE) in the United States defines its purpose as 'to promote the development of collaboration, and inter-disciplinary understanding of human ecology and its application'. 'Providing a forum for the exchange of information, the advancement of inter-disciplinary studies, and the identification of problems and their possible solutions' is mentioned as its specific objective (SHE, 1991).

The Latin American Centre for Social Ecology (CLAES), based in Uruguay, looks:

> for an alternative and co-participative development in Latin America that links its social and environmental dimensions. It is a development which is based on a consideration of BEINGS, all beings, human or not; but not on GOODS. In one of its programs CLAES suggests that it is directed to people living in slums, on the outskirts of city areas. It is specially concerned with marginalised groups: particularly the poor, women, children, the disabled, the chronically ill.
>
> (CLAES, 1991)

This highlights clearly the differences of focus among human ecologists. Such differences define their priorities and confirm their attitudes to all life, and to human life in particular.

After Rio

Human ecology during the first half of the 1990s was concerned with health, sustainable development, the widening gap between 'South' and 'North', and the increasing distance between rich and poor, both in the 'South' and in the 'North'.

When human health is placed at the centre of frameworks for development, social justice becomes a yardstick for appropriate sustainability. Hence, human ecology becomes more focused on issues relating to the health of individuals in healthy environments.

We will return to these issues relating to health and human ecology later in this chapter.

Definitions

Many attempts have been made to define human ecology; these efforts continue. Although there have been changes of emphasis in the central themes and concerns of human ecology, the foundations and principles remain unaltered. One of the significant changes of emphasis is that in earlier definitions there was an emphasis on 'investigation and study', while in more recent definitions considerations of 'problems and problem solving' have become more prominent.

We have seen that Park, a sociologist, coined the term human ecology. In one of his papers on the subject, he summarises his views by defining human ecology as 'an attempt to apply to the inter-relations of human beings a type of analysis previously applied to the inter-relationships of plants and animals'. Park concluded his definition by placing emphasis on the state of balance and equilibrium: 'Human ecology is fundamentally an attempt to investigate the process by which the biotic balance and social equilibrium are maintained once they are achieved and the process by which, when the biotic balance and social equilibrium are disturbed, the transition is made from one relatively stable order to another' (Park, 1936). We call this equanimity.

Barrow, a geographer, in 1923 made a clearer reference to the environment. He suggested that: 'Geography is the science of human ecology, as geography will aim to make clear the relationships existing between natural environments and the distribution and activities of man'.

Philosophical approaches also find a comfortable place in human ecology as they have the capacity to link human values and human potential (Westeny, 1988). Westeny identifies self-actualisation as resulting from efficient and effective interaction between individuals and their environment.

Psychology has been interested to apply an ecological framework in order to identify and examine the role of non-psychological inputs (i.e. the larger ecological environment) into human behaviour (Young, 1974).

Green (1969) claims that: 'economics is human ecology; it is the study of man's adaptation to and creation of an economic environment resulting from those forces that maintain a dynamic society'.

Straus proposes a discipline which bridges the gaps and deals with complexities. He suggests: 'Human ecology seeks to understand and manage wisely the complex problems of the planet of which humans are a part. It integrates the old disciplines of highly specialised scientific investigation with the new discipline of seeing things, and acting upon them as generalists' (Straus, 1990).

These definitional examples highlight the aim of most human ecologists who wish to provide a framework for living wisely and enhancing specific capabilities; human ecology as seen in its broadest definition is concerned with the ways in which humans use resources. Such a framework is based on efforts to understand social organisations and the inter-relationships between humans and their physical environment; a concept adapted from ecology. The term 'environment' – as used in some definitions – denotes the natural or social surroundings of humans, more completely both.

Another claim is that human ecology considers 'conditions of existence' produced and influenced by human beings. From this forum, therefore, the most fundamental questions can be asked.

The World Health Organisation European Office has attempted a comprehensive definition of human ecology:

> A holistic and integrative interpretation of those processes, products, orders and meditating factors that regulate natural and human ecosystems at all scales of the earth's surface and atmosphere. . . . It implies a systemic framework for the analysis and comprehension of three logics and the inter-relations between their constituents using a temporal perspective.
>
> These three logics are: bio-logic, or the order of biological organisms; an Eco-logic, or the order of inorganic constituents (e.g. water, air, soil and sun); and a human-logic, or the ordering of cultural, societal and individual human factors. It is suggested that this macro-system of three logics regulates the world.
>
> (WHO-EU, 1993: 222)

Gerald Young, who has devoted his life to writing and advancing human ecology, gives a detailed account of different aspects of it. He examines various characteristics of human ecology which he has constructed around these dimensions.

- Human ecology is inter-disciplinary. It exists on the margins of each specific social science. With their commonalities analysed and integrated, human ecology becomes the incorporative subject for the study of humankind.
- Human ecology is connective: it has a leadership role in the search for connections between micro and macro levels of scholarship.
- Human ecology is committed to synthesis, which is an attempt to understand or combine and unify isolated data and make sense of them in a recognisable whole.
- Human ecology is transcendental, which is to realise a greater truth and go beyond the limits of traditional disciplines; it should transcend narrow specialisation.
- Human ecology eschews chauvinism. Human beings face a multitude of difficulties in their relationships, first, to themselves and those of their kind; second to their neighbours and other kinds; and third, to the earth or natural environment. Human ecology is able to use all the inputs and varieties of perspective it can command.
- Human ecology is holistic. It accommodates the full range of human interactions; and is concerned with all the levels of complexity that humans encounter from the sub-atomic to the ultimate expanse of the universe.

- Human ecology is humanistic. Humanism has traditionally been an educational tool in the creation of human community by inculcating common ideas and by stressing reciprocal responsibilities among human beings.
- Human ecology is, by definition, anthropocentric. To humans it marks the ecology of ultimate concern.
- Human ecology concedes a subjective reality, recognises the role of emotions, the contribution of intuition, the surprise of serendipity, and the unique collection of inheritance and experience gathered together in each individual human being.
- Human ecology is process oriented. Ecology studies process. Human ecology is the study of relationships, not things.
- Human ecology does not deny teleology but human ecology remains committed to functionalist statements and how things work. The human systems that we work with are goal directed. So human ecology does not embrace teleology, but does not deny it either. In fact it considers all possibilities.
- Human ecology acknowledges mystical or spiritual dimensions. It is a science that tries to understand people, their thinking and behaviours; hence spiritual becomes part of the focus.
- Human ecology denies determinism. It cannot be deterministic, but is always relativistic.
- Human ecology seeks to understand community. Community is one of the most fundamental units in biological ecology. Humans are a social species. It raises questions about the role of 'individuals'.
- Human ecology recognises some form of family as a fundamental ecological unit. Family is not recognised in biological ecology. Family would fit between individuals and populations; a unit of undeniable significance in the formation of individual humans; the basic unit of material consumption in the human community; a buffer between individuals and larger groups; the source of identity and succour for human beings.
- Human ecology recognises the wisdom of vernacular, to examine and then adapt the way that the same concepts are utilised in everyday life.
- Human ecology includes a normative vision.
- Human ecology is subversive.

(Young, 1991)

Most studies in recent years have focused on the capacity of human ecology to cover a broad area which is concerned with humans and their environment from a trans-disciplinary point of view.

Here I attempt to capture the essence of human ecology by listing its dimensions rather than attempting to establish acceptable definitions. Human ecology from our perspective:

- is concerned with causes and effects, events and impacts, and relationships and inter-dependence of humans and their environment;
- is concerned with global, spatial, temporal dimensions and their interplay;

- focuses on people and their material, cultural and spiritual needs, and their interactions;
- emphasises that development must be appropriate, not only to the environment and resources but also to the culture, history, institutions and social systems of the place in which it occurs;
- integrates knowledge, experience and wisdom from all disciplines, sciences and theories in order to link past, present and future together; and
- is problem oriented and solution focused.

Core concepts

There is a diversity of views concerning the definitions and dimensions of human ecology. Here is a overview of some of its main concepts.

Sontag divides the core concepts of human ecology into two main categories: 'general systems concepts' and 'ecosystem concepts'.

1. *General systems concepts* include input, throughput, output, feedback, boundaries, interfaces, open and closed systems.
2. *Ecosystem concepts* comprise

   - *Environmental concepts.* Food, clothing, shelter, form, use, design, aesthetic qualities, meaning, adequacy, resources, matter, energy, information, socio-historic-cultural patterns, reference groups, significant others.
   - *Individual and family concepts.* Human development, human needs, family functions and structure, values, goals, transformation processes, family roles and transitions, individual life transition, resources, decision making.
   - *Interaction concepts.* Inter-dependence, functional relationship, perception and response, adaptation, communication, stress, conflict, information processing, management, space–time relationship.
   - *Outcome concepts.* Quality of life, environmental quality, self-formation, social goals.

                                                                  (Sontag, 1988: 126)

Westney categorises the core concepts in human ecology differently. He suggests three concept groups which relate to human beings, to the environment, and to the interactional outcomes of human beings.

Concepts concerning human beings, including:

- the nature of individuals, their needs and their effort towards development;
- the nature of human beings (i.e. a composite of physical, social, intellectual, emotional, and spiritual components);
- human potential as reflected in life span development and essential human needs (i.e. food, clothing, shelter, safety, security, friendship and affection, need for respect and esteem, aesthetic and self-actualisation);

- developmental tasks, human resources, human capital, human interaction, human condition, human rights, human wellness, human well-being, prevention of human problems, and human services.

Concepts concerning the environment of human beings including:

- the micro environment (family, home and communication with others); and
- the macro environment (neighbourhood, community, culture, educational system, and economic system, as well as water, land and pollutants).

Concepts concerning interactional outcomes of human beings. Key concepts here include:

- quality of life, which is the degree to which needs are met and the level of satisfaction one experiences in the process of life;
- management of self and life situations (skill in administering physical, intellectual, social, emotional, moral, economic and spiritual life in order to develop one's life situation);
- the fully self-functioning person. Someone who develops his or her potential and meets his or her needs consistent with available resources as well as ability to contribute to others.

(Westney, 1988: 132–135)

Holistic human ecology

During the past few decades excessive specialisation in various disciplines has taken place. Guardians of particular disciplines both individually, and more formally in their annual association meetings, have re-enforced their own boundaries and defence lines. The situation in many ways is similar to the feeling of *nationalism* in newly independent countries.

This application of the new *disciplinism* was taking shape and taking place in the form of technology, progress and development, while connection or communication between disciplines was diminishing. Attention to general aims and ultimate expectations was being neglected.

It has taken us only a few short decades to begin to comprehend the negative impacts of over-specialisation on human health and the environment.

Larger questions began to be asked in the early 1970s. The validity and value of many so-called scientific successes came under scrutiny. The absolutism of many disciplines – such as medicine with its claims to be able to 'fix' anything – was questioned in some Western societies, while in poorer countries such absolutist arrogance continued to prevail.

The need for a holistic approach has gained momentum in recent years, thus ensuring that the more fundamental questions are being considered. Different disciplines and professions find comfort in the use of the term human ecology which enables them to express something beyond the limited view of their own discipline.

The role of a holistic approach in human ecology has been justified in the following ways:

* it has its roots in pure science;
* it has flourished in the social sciences;
* it is responsive to contemporary issues; and
* it has practical as well as theoretical applications.

The ecology of human beings is concerned with connections. It has given itself the task of bringing together many disciplines in order to enhance the human condition and the quality of life. The nature of any true intellectual challenge must be based on an aspiration towards integration.

Human ecology provides a bridge between specialisation and generalisation, between integration and fragmentation, between academic and ordinary life, between the past and future, and between individual and global issues and approaches. One of the main dimensions of human ecology has been its holistic framework.

Carpenter (1988) suggested that human ecology, by seeing individual events in their total ecological context, facilitates a correction of this fragmentation of knowledge and thus re-integrates these compartments of human perception.

There are many reasons why the descendants of discipline-based specialisation find a sanctuary in human ecology. The holistic framework of human ecology:

* challenges the status quo;
* enables foci to range from the micro to the macro spectrum;
* provides a forum for questioning the ultimate outcomes of specialisation;
* forms a bridge between the intellectual and community concerns;
* interprets processes, products, ordered systems and meditating factors that regulate natural and human ecosystems;
* draws on and re-conceptualises material from other disciplines as required in the expansion of questions and the solving of problems; and
* challenges concepts of life and the nature of totality.

Zena Daysh provides a historical overview and connects the holistic approach to the fundamental idea of sustainability:

> In the first cultural human, there is a deism in nature. The root of this sense is inter-dependence which is central to ancient eastern philosophies. People were unified by their philosophy, spiritual and intellectual, and their behavioural patterns, that intuitively, in a consciousness not fully comprehended, helped them to live with the principles of co-existence, and what today we would call SUSTAINABILITY. Those early populations could have no doubt that they were an integral part of nature, living as they did in greater intimacy, and for that matter, more immediate conflict with nature. Those ancient times could truly be looked upon as a kind of human

ecological way of life – an ecosystem, where balance was the recognised necessity.

(Daysh, 1988: 20)

While the holistic approach in human ecology has been a matter of emphasis, its professional dimension has not been challenged. Straus commented on this dual role of specialist and generalist in human ecologists, who are:

> sometimes contributing the expertise gained in micro-focused research into the far frontiers of new knowledge; at other times collaborating in the task of selecting, evaluating, integrating and assembling essential bits of knowledge into a micro-matrix of the issue under discussion. Only a few become comfortable and skilled in both the specialising and generalising functions.
>
> (Straus, 1990)

Indeed, this is its essential promise and challenge.

Human ecology creates an area for its practice of generalising – combining old arts and new skills so as to reach decisions in a way that seeks to include all relevant knowledge and all concerned points of view. Again because of an absolute necessity, the inter-disciplinary, holistic and multi-dimensional approach has become a common desire among many intellectuals: 'What we need today is to relate the sciences to each other, and we lack the specialist in generalisation to do this' (Pagel, 1988: 36).

This dual capacity of human ecology is rooted in its concern with all significant aspects of reality. It is the specialisation in generalisation that recognises that various disciplines should discuss the usefulness of their approaches to the ecology of human situations.

Increasingly, human ecology is being acknowledged as the orientation or consciousness which will assist in bringing us through present crises and in promoting sustainable development for the future.

The holistic view of human ecology ensures that the dynamism, continuity, totality, harmony and productivity of human ecosystems are maintained – human ecosystems in this approach being concerned with the flow of cultures, covering all human activities, through natural interactions; and human ecologists, being specialists in integration and non-specialisation, contribute to the advancement of individual, societal and global well-being, and hence to life's qualities.

Values

Human ecology has a special place in both arts and sciences. It is rooted in 'pure' science and the social sciences, and has a niche in the world of arts through the trans-dimensional characteristic of its subject matter – human beings.

The category of liberal arts is based on values, and at the same time creates values:

Its purpose is a wholeness of vision achieved through an intuitive transformation of subject into form. What the novelist or composer does is approach complex problems of relation in the world by transforming them into problems of a given artistic medium where the aesthetic skill and training allow them to be solved.

(Carpenter, 1988: 1–2)

Human ecology provides a vehicle for re-engendering and confirming a faith in human beings – that we are able to solve problems in the world. Most of these problems are processes, not 'faits accomplis'. This emphasis on values in human ecology enables us to ask 'big questions', challenge solutions and transform processes into objectives. Focusing on objectives and procedures, macro and micro situations, the inner and external worlds of humans are all well built into values of human ecology.

Human ecology seeks to clarify those complex issues which are of concern to human beings in relation to the environment. At the heart of every complexity lies the inseparable concept of 'justice and equity'. Opportunities for equitable use of resources at local, regional and global levels are the core concepts in this regard.

If human ecology is to give a lead in our day – and surely we must claim that it can and should – we need a cohesive view of history; a cohesive view of the co-evolution of humankind and the world of nature. We need to understand the development of humans and their societal institutions within the parallel evolution of nature itself.

Human beings are the most singular creatures in our known universe: their ability to contemplate life is fundamental. Life is a right, not a social, political or economic issue. Whatever our cultural or religious orientation, we live because we are alive, because we *can* contemplate life. Asking questions about life is the most common experience among individuals; it is much more than, and beyond, survival.

Destruction of the environment has put not only the external environment but also the interior human mind, body and psyche in danger. Ecosystems function well without human beings but human beings cannot survive without ecosystems. Human adaptation to the environment is cultural, not biotic. It is thus the adaptation of culture to the utilisation of resources which finds prominence in human ecology.

The simple fact of being alive in a viable society is a core value of a sustainable society. A sustainable society 'gives life core value; affirms love as a prime value rather than domination and dominion; affirms justice and security as other prime values, and encourages self realisation as a key to a fulfilling life' (Milbarth, 1991: 3).

The behaviour of individual human beings is far from predicable, and almost all societal trends are based on the behaviour of individuals. Human beings, individually, are at the very centre of human ecosystems. Human ecosystems illustrate their life cycles, at individual, family, community and global levels. Study of the historical development of human ecology provides the aptitude to approach the macro and micro ecosystems (individual, societal and global) from a holistic standpoint.

One of the principal ideologies in human ecology is that we can achieve a healthy life in a healthy environment if we live in harmony with other human beings and with nature; however, we need a far more sophisticated means of ethical, cultural and technological judgement.

The following is a list of values around which the knowledge and practice of human ecology can integrate. Such a list may be ordered differently in various settings:

- *Equity* and justice in social relations.
- *Access* to opportunities for human development and resources.
- *Respect* and caring for the survival, worth and dignity of all people.
- *Respect* for the conservation, worth and dignity of the environment.
- *The co-existence* of humans with nature.
- *Work* as a means to the fulfilment of human existence and survival.
- *Aesthetic* values such as beauty of environment.
- The core concept of *health* and *well-being* for all.
- An emphasis upon *knowledge* and learning.
- Support of and for individual *rights* and dignity.
- Acceptance of all *cultures* as worthy of respect.
- *Peace* at a personal, community and global level.

Applications

There has been much emphasis on the applied aspects of human ecology since the 1920s. Various levels of application have been suggested by professionals from different fields. Human ecology promises a different arena in the identification and understanding of problems, problem solving and strategies for change. The central use of human ecology is to identify the place, process and presence of people in nature.

When Park and McKenzie published their paper in the book *City*, in the 1920s, private companies rushed in, to order research on the future of the physical growth of cities in order to enhance the profitability of their investment plans.

Westney sees human ecology as an applied science which 'seeks to identify the forces which enhance human development, actualise human potential, optimise human functioning, and improve the human condition and quality of lives of people' (Westney, 1988: 129).

Johnson, in an address to a group of human ecologists, told them:

> The world is ready to hear your message. The world is faced with such serious problems that we must respond. We must bring back health-enhancing, not health-threatening water. We have to come up with solutions for housing, transportation, and energy efficiency; positive results in these areas could

bring great gains to global society. We have to communicate and define our idea in a way that the public could react to it.

(Johnson, 1990)

We should be able to identify a responsibility for human ecologists and the application of their theories, framework, methodologies and knowledge in any arena of human interest.

It might be useful here to outline examples of such roles in dealing with issues such as challenging complex processes, decision making, problem solving and curriculum development.

Most problems of the contemporary world have roots in both natural and cultural factors. The orientation of human ecology to natural and cultural ecosystems is capable of integrating and guiding our quest for identifying key problems for solution, in order to develop coherent strategies for change.

The usefulness of human ecologists in problem solving can be identified as:

- placing a more conscious focus on process;
- making more effective use of resources;
- integrating analysis with intuition;
- exploring the broader impacts of new initiatives;
- recognising the behaviours and reactions of individuals and groups; and
- taking account of and understanding people's views and reactions.

A basic understanding of human ecology is recommended for students, professionals, politicians, and decision makers, as well as for individuals more generally.

Such an educational programme on human ecology would include:

- a global ecological view;
- qualitative and quantitative conceptions of populations and resources;
- a recognition of the multi-dimensional underpinning of knowledge drawn from the actualisation of public policies;
- the history of ecological and technological relations;
- the present and future directions of biological and cultural diversity;
- a fundamental knowledge of ecosystems, both natural and human;
- a consideration of applied human ecology including ethical education and learning; and
- problem solving and imaginative alternatives for the future.

Levels of human ecosystems

Human ecology includes the study of *human ecosystems*, however, the dimensions and extent of human ecosystems are not easy to define, and depend on the background of those who are engaged in the subject.

Human ecosystems are complex, more so than physical ecosystems. By drawing upon the concept of ecosystems in studying human life conditions, we are acknowledging such immense complexities, and inter-dependencies of all the various organs and members in any natural setting.

Human ecosystems are the most viable frame in which the inter-dependence of individuals, families, communities and the world can be observed and examined. Sontag explains this with clarity. He says:

> Human ecology extended its view to incorporate the global inter-dependence of individuals, families and communities with the resources of natural, constructed and behavioural environments for the purpose of wise decision making and the use of resources essential to human development.
>
> (Sontag, 1988: 119)

Westney views human ecosystems according to their input–output factors. He considers the input factors as human beings and their environment, and the output factors as human beings at any point of the developmental process. He suggests that the organism is endowed with innate potential; and when optimally actualised, this results in a fully functioning person (Westney, 1988: 131).

Examples of human ecosystems can be found anywhere: in the setting of a spring, from which a woman draws water for cooking and washing and hygiene; or in kitchens of households with their great variations in different cultures; so too in the Hong Kong stock exchange.

The following are different recognisable levels of human ecosystems.

Individuals

Human ecology is the study of humankind in ordinary daily life. It deals with everyday processes of living. A fundamental promise of human ecology is to offer a way to develop the capacity to manage daily lives in a balanced and efficient manner.

The concept of individuals as units of a human ecosystem allows fundamental issues such as health of individuals, rights of individuals, and equity and equality of individuals to be considered in relation to resource use, culture and technology.

There is much debate in many disciplines about the role and responsibilities of individuals in the safeguarding of rainforests, alluvial fans, mangroves, or historical sites and monuments. There have also been explorations of the responsibility of individuals towards themselves, their societies and the whole universe; so too concerning the roles of individuals in health, human rights, in equitable allocation of resources and in applications of technology.

Human ecology, by providing a framework for a better understanding of these issues and of human–environment interactions, contributes to individual well-being. The ultimate aim is the creation and maintenance of healthy people in healthy environments.

This basic level of human ecosystem focuses on the physical, mental, emotional, developmental, societal, economic, political and cultural interactions of individuals in their environments, both natural and cultural.

Families

The family, as the second stratum of human ecosystems, is unique to human beings in contrast to natural ecosystems. It consists of a population of human beings bound together and depending on each other. It is the environment least distant from individuals in interaction with the other elements of life and living. The family is a buffer against the hostile outside world and is a setting for actualising the development of the potential of individuals in preparation for their life path.

Human ecology sees the family as the fundamental unit of human ecosystems. This is where human ecosystems differ from others. The family is where members are prepared for life, both individual and societal. It is where essential resources are available, where people learn how to use resources, where patterns of human behaviour are established and grown, and where links between people and their environment are built.

Sontag (1988) suggests that families, both as units and as social institutions, build and maintain systems of interaction which lead to the maturing of the self-formation of individuals and to enlightened cooperative participation in the critique and formulation of social goals as well as the means for accomplishing them.

Bubloz defines families as:

> one of the components of the inter-dependent and inter-acting set of systems that comprises community, societal, and global human ecosystems, that is macro-systems. Families represent one of the primary ways by which a population of inter-dependent organisms, each individually not equipped to survive alone, organise to meet a variety of goals and diverse environmental conditions as they change and evolve over time.
>
> (Bubloz, 1991: 26)

The family is also linked to interactional concepts which include inter-dependence, functional relationships, perception and response, adaptation, communication, stress, conflict, information processing, management, and space–time relationships. It also includes other concepts such as quality of life, environmental quality, self-formation, and individual and cultural goals.

The concept of the family is of great importance in studying human ecology. It includes issues such as human development, human needs, family function and structure, family values and goals, resource use and decision making. Whatever decisions families make affect the environment.

Communities

The community is where all elements, factors and organisms meet, interact, and interplay, all affecting one another. The natural community includes everything in action, at any given place at any given time. Human communities are different because of the complexities of human interactions and behaviours which includes physical, mental and spiritual needs.

Communities of other species are bound together by energy flows and the cycle of materials. Human communities are bound together by higher levels of need; emotional needs, spiritual needs, a desire for enhancement of life, curiosity, and a sense of responsibility.

The term *culture* has the capacity to capture all such needs and aspirations. Without culture, a human community is an empty shell, a vacuum. Communities with cultures based on discrimination – whether racial, professional or financial – degrade nature and humanity.

The dignity of human beings and nature can only be maintained in those communities with equity and respect for all members, whether humans or trees or ants. As expressed by Abu-Sa'id Abul-Kheyr, tenth-century Persian Sufi Master and poet: '*God created you free, be free.*'

The world

Human beings, tiny elements on a universal scale, can contemplate the whole universe. They can consider the totality of nature and the nature of totality. The largest and uttermost extent of human ecosystems is, therefore, the world within us, the world around us, and the world beyond us.

Concerning the universal aspect of human ecosystems, everything is connected to everything else. There is a strong notion of relatedness and connectedness everywhere which spans time and place. Against this background some themes come to be of central importance: balance in nature, reverence for life and social justice for all.

Health

The most valuable things in life are the simplest. *Being* is the most valuable and the simplest. If thinking defines *being* ('I think, therefore I am'), it is the *feeling* which defines *being healthy*.

Health can be defined as the status of 'being'; how we feel inside ourselves, and how we are seen from the outside, at a cosmic level and in comparison to others. *Being* encompasses the totality of our relationships with ourselves and with all else. Therefore the health of human beings is contained in the nature of relationships to whatever surrounds them; the environment as totality – all there is.

Sometimes the term *health* is used as a tool and an indicator to describe the quality of relationships between individuals, their communities, and their environment.

I am therefore identifying a deeply embedded dual function for the term *health*: as an expression of the status of being, and as an indicator of relationships with environment.

A human ecological perspective – as a crossroad between cultural, social, economic, physical and biological concepts – is capable of comprehending, framing and actualising such a dual function for health. Reductionist professionals on the other hand, especially those functioning in a bio-medical construct – who seek to control the human being body for profit – are not capable of such a comprehension.

The human ecological approach should be used as a tool to question the manipulation of the health-illness agenda by medical, economic and political groups. The term *health* in this regard covers a wide spectrum, including health policy making, health research and development, health care delivery, as well as defining the health status of individuals, families, communities and the globe.

Solorzano came to the same conclusion in 1993. He suggested: 'While the ecological perspective maintains that the alleviation of diseases, the creation of happiness, and well-being takes priority over other incentives; twentieth century biomedical medicine seems motivated by profits, political gains and scientific reputation' (1993: 287).

Health becomes synonymous with wellness, happiness and satisfaction with life and being. It is a degree of physical, emotional, intellectual, social, moral and spiritual robustness which characterises the individual at any point in time. Health of human beings is that state of satisfaction or happiness – well-being – which reflects the degree of perceived fulfilment of physical, psychological, spiritual, social, aesthetic and material needs and wants.

Wellness can only be defined in relation to cultures. Illness too can be defined in relation to cultures. If wellness and illness are regarded as bio-medical concepts, human and cultural concerns are ignored. Understanding and promoting health can only be achieved through a cultural view of both wellness and illness.

Hence, *health*, in my view, is *a sustainable state of well-being, within sustainable ecosystems, within a sustainable biosphere*. That is a definition that I feel comfortable with. It strongly emphasises positivity, without diminishing the negativity of its dimensions.

The health of any species cannot be separated from its ecosystem. Consequently health may also be examined at those same levels considered in the section on human ecosystems.

Individuals

For an individual being healthy means being *physically sound, mentally intact, spiritually happy, socially active, politically aware, economically productive and culturally responsible*.

This list implies a degree of balance in human life factors, which goes far beyond biological life. It emphasises that state of balance or equanimity as being normal, rather than being perfect. This definition is able to include all aspects of disability or chronic illness. A person lacking cultural responsibility could be defined as more unhealthy than someone without an organ, whether visible or internal; without a hand or a kidney.

To individuals, being healthy means a stable and enhancing quality of life; a happy personal, family and social existence; and the opportunity to make choices.

The WHO European Regional Office suggests a list of prerequisites for health which include: freedom from war, equal opportunity for all, satisfaction of basic needs (food, education, clean water and sanitation, and decent housing), secure work and a useful social role, political will and public support. It is obvious, yet worth mentioning that dimensions, elements and factors have different weightings at various stages of life.

By this definition, the health of individuals can be sustained in healthy homes and healthy environments.

Communities

To communities, health means equality; equal opportunities for people to live and die with dignity and use natural and human resources appropriately. Health at a community level means freedom from war, discrimination, disease, hunger, fear, illiteracy, ignorance, pollution, homelessness, profiteers, and lies.[2] A healthy community provides its people with satisfaction of basic needs (food, housing, educations, hygiene, and care)[3]; secure work and employment; and social and cultural support and appropriate roles.

The health of communities can be ensured by maintaining a balance between societal responsibilities and individual rights. A community is responsible for providing equal personal, educational, social, cultural, economic and political opportunities for its people in order to actualise their potentials. Individuals, on the other hand, utilise such opportunities to enhance the quality of life of themselves and others, as well as to enhance the spirit of community. Healthy communities sustain equal opportunities for utilisation of cultural and geographical resources. And finally, it is the community which should decide what is healthy and what is unhealthy.

The world

The health of human beings, or any species, whether as individuals, families, communities, or populations, is ultimately dependent on the health of the whole world. The health of whole systems, or the whole world, is only maintained through a delicate balance between the needs of people – physical, biological, spiritual, mental, cultural, economic and political – and the satisfaction of such needs – by the appropriate utilisation of resources – both environmental and human.

Human ecology provides a holistic model for the maintenance of such balance: by the fulfilment of needs – food, clothing, shelter, security, and care – and the utilisation of resources. This framework integrates individuals, society and governments, and emphasises the responsibilities of each.

Thus at a global level, health ultimately means equity; an understanding between cultures; and a co-existence free of colonising attitudes, fear, threats or war.

I would like to restate my definition:

health is defined here as a sustainable state of total well-being, within sustainable ecosystems within a sustainable biosphere. The creation and maintenance of healthy homes, within healthy communities, within a healthy world is the major objective of human ecology.

Health ecology

Health is a core concept of human ecology. Conversely, human ecology is the key to a comprehensive understanding of health. There have been many intellectual activities concerning health which have drawn on human ecological approaches over the past two decades. This reflects the increasing importance of 'healthy environments' and a concern for human health and conservation. It reflects the need for changing attitudes and for development of behaviours and policies in order to maintain the health of individuals, communities and the globe.

The Rio Declaration reflects this: 'human beings are at the centre of concerns for sustainable development. They are entitled to a healthy and productive life in harmony with nature' (UNCED, 1992).

The importance of a human ecological approach to health is integrative, holistic and positive. It views health as an inseparable part of life – a life itself based on the wisdom of a healthy life; a healthy life which is concerned with all its people and resources, a healthy life which is creative.

There has been increasing support for the re-alignment of newer health ideas with human ecology, and for the belief that human ecology should place health at the core of its principles and approaches.

I wish to emphasise that health and environment are trans-disciplinary, multi-dimensional and holistic in nature. It is their inter-relationship, inter-dependence and interplay that should be the subject of intellectual, managerial and community concerns. These concerns have created the necessity for new thinking, vision and actions.

New approaches are required to solve non-linear complex issues. The central aim of this book is to exemplify and respond to such needs at the theoretical and the applied level.

The human ecological perspective is appropriate because human ecology adapts its sources of information and methods to suit the dynamically changing nature of the complex systems in which people wish to develop a view.

The term 'health ecology' represents a human ecological approach to health and well-being. Health ecology is concerned with a comprehensive picture of human health at individual, societal and global levels. It is used to illustrate patterns of human health in relation to the environment; physical and social, micro and macro, cultural and technical, individual and universal.

The key issues in health ecology are:

*   *The multi-dimensional nature of the environment.*
*   *The integrative relations that humans develop in inter-action with their environment.*

- *Sustainability*, which is the recognition of fundamental limits in utilisation of resources. It is necessary to re-examine paradigms, principles and practices of sustainability in relation to human health.
- *Life expectancy* seen not only in terms of longevity, but also as concerned with expectancy of development at different stages and in different settings, establishing the elements of quality in life.
- *Health as the core element of the public domain* concerned with community participation, advocacy and empowerment.
- *The provision of a theoretical model which offers a logical framework for interactions, at various levels,* of individuals, institutions and open communities.
- *The assessment and management of resources and risks in and for the creation of healthy ecosystems.*

The concept of health ecology also provides a framework for problem solving and developing strategies for change, which includes:

- a holistic vision for social, environmental and health planning and strategy development;
- an ecological framework for policy planning and strategy development for families, children, and special groups;
- the advocacy, planning and strategy development for healthy communities;
- a theoretical model which offers a logical framework for the appreciation of environmental issues at various levels;
- an assessment of the health of individuals and communities, which demonstrates the direction and measures the success of social and environmental policies;
- the ability to relate the state of the health of individuals and of the community to national goals and objectives;
- the provision of health inputs to environmental management teams, which ensures the inclusion of all relevant tangible and intangible factors in any programme procedure, implementation or evaluation;
- comments from a 'health' perspective on development proposals;
- contributions to Environmental Impact Assessment from a health point of view;
- contributions to site selection for large scale projects;
- the ability to establish criteria and evaluate residential, industrial and functional units; and
- the facilitation of community participation in planning and evaluation processes regarding community health concerns.

Health impact assessment

Health is the main resource for any type of development: conceptual, cultural, societal, environmental, educational, structural, health–illness care delivery, technical, economic, political or technological.

Creation of a healthy existence and enhancement of life qualities for individuals, families, communities and the world, have been highlighted as the main objectives of health ecology throughout this chapter. Health impact assessment is a formal tool to achieve such aims, and the application of health ecology towards such aims.

Health impact assessment requires much more work, and further developments. The final chapter of this book highlights important technical issues in this regard, but the application of it to the health and the living conditions of the Third World needs more practical and innovative actions.

Health ecology: a gathering of perspectives

I don't know when the idea of *health ecology* occurred to me; if it helps I can say: I was born with it, and I have grown with it.

In my life, I have been involved in a wide range of disciplines, including history, philosophy, human and economic geography, anthropology, cultural studies, environmental studies, public health and environmental health, water supply, water and land-use, folklore and literature. They have all been necessary, but not sufficient to the understanding of health in ecosystems. My strength has been to remain at the core of a trans-disciplinary approach.

In 1988 I presented a paper on health impact assessment and the use of health indicators, as applied to health ecology. In 1990, my colleagues and I developed and taught the subject of health ecology on a health studies course at the Faculty of Health Sciences at the University of Newcastle.

During a study leave in 1994, in the UK, I started to put together my ideas to prepare a framework for health ecology. Soon I realised the importance of bringing together different views and the need to gather the perspectives of people from different backgrounds. This book is the outcome.

Notes

1 I presented a paper to the International Association for Impact Assessment in Brisbane, Australia, in June 1988, while working at the Australian Institute of Health. It was titled 'Health Indicators and Health Impact Assessment'. I highlighted some basic definitions for health ecology, and introduced Health Impact Assessment as an applicable methodology for health ecology.
2 Darius, the Achaemanid King of Persia, in one of his inscriptions in Persepolis, says 'May Ahura-Mazda protect this country from a hostile army, from famine, from the Lies. Upon this country may there not come an army, nor famine, nor Lie.' (Ghirshman, 1954, p. 154).
3 There are many caring facilities that should be provided by the community for its people; health care delivery is one of them, welfare is another. Health care delivery, at any level, should participate in and not control health and health policies.

References

Adams, Charles C., 1935, The Relationship of General Ecology to Human Ecology, *Ecology*, 16, July: 316–35.

Aggleton, P., 1990, *Health*, Routledge, London.

Alexander, L., 1994, *Health Related Indicators for Sustainable Community* (paper), Department of Geography, Lancaster University.

Alihan, Milla Aissa, 1938, *Social Ecology: A Critical Analysis*, Columbia University Press, New York.

Andrews, M.P., Bubolz, M.M. and Paolucci, B., 1980, An Ecological Approach to Study of the Family, *Marriage and Family Review*, 3: 29–49.

Anon., 1990, Human ecology and environmental politics. *Political Geography*, 9: 103–107.

Anon., 1993, Quality of life as a new public health measure, *Morbidity and Mortality Weekly Report*, 43(20): 375–380.

Ashton, J., and Howard, S., 1988, *The New Public Health*, Open University Press, UK.

Ashton, J., 1992, *Healthy Cities*, Open University Press, UK.

Ashton, J., and Seymour, H., 1993, Setting for a New Public Health, in: Beattie, A. *et al.* (eds), *Health and Wellbeing: a Reader*, Macmillan, UK.

Australian Institute of Health, 1990, *Australia's Health*, Australian Government Publication Service, Canberra.

Australian Institute of Health, 1992, *Australia's Health*, AGPS, Canberra.

Australian Institute of Health and Welfare, 1994, *Australia's Health*, AGPS, Canberra.

Baric, L., 1991, *Health Promotion and Health Education*, Barns Publication, UK.

Barrow, H.H., 1923, Geography as human ecology, repr. in: Young, G. (ed.), 1983, *Origins of Human Ecology*, Hutchinson, Stroudsburg.

Baum, F.E., 1993, Healthy cities and change: social movement or bureaucratic tool, *Health Promotion International*, 8(1): 31–40.

Bernis, C., 1991, Global changes and their implications for women, *Journal of Human Ecology*, 2(1–2): 171–196.

Bergstrom, S.K.D. and Ramalingaswam, V., 1992, Health, in: *An Agenda of Science for Environment and Development in the 21st Century*, Cambridge University Press, based on a conference held in Vienna, November 1991.

Better Health Commission, 1986, *Looking Forward to Better Health*, AGPS, Canberra.

Blaxter, M., 1990, *Health and Lifestyles*, Routledge, London.

Bogdan, J.L., 1984, Family organisation as an ecology of ideas: an alternative to the ratification of family systems, *Family Process*, 23: 375–388.

Borden, R.J. and Jacobs, J., 1988, *Human Ecology: Research and Applications*, Society for Human Ecology.

Bridgewater, P., 1992, *Biodiversity: Broadening the Debate, A Trilogy of Discussion Papers*, Australian National Parks and Wildlife Services, AGPS, Canberra.

British Medical Association, 1987, *Living with Risk*, Wiley Medical Publications.

Bronfenbrenner, U., 1979, *Ecology of Human Development: Experiments by Nature and Design*, Harvard University Press.

Bronfenbrenner, U., 1986, Ecology of the family as a context for human development: Research perspective, *Developmental Psychology*, 22: 723–742.

Brown, M. and Paulucci, B., 1978, *Home Economics: A Definition*, Washington DC, American Home Economics Association.

Brown S., 1988, The Unequal Distribution of Death and Illness, *New Doctor*, No. 47.

Bubolz, M.M., Eicher, J.B. and Sontag, M.S., 1979, The Human Ecosystems, *Journal of Home Economics*, 71: 28–31.

Bubolz, M.M. and Sontag, M.S., 1988, Integration in Home Economics and Human Ecology, *Journal of Consumer Studies and Home Economics*, 12: 1–14.

Burau, V.D., 1994, *Entering the Secret Garden of Medical Autonomy: State Regulation of Gatekeepers in Britain and Germany*, Paper prepared for workshop on the State and the Health Care System, 17–22 April.

Caplow, T., 1949, The social ecology of Guatemala City, *Social Forces*, 28, pp. 113–135.

Carpenter, W., 1988, Human Ecology: The possibility of an Aesthetic Science, in: Borden, R.J. and Jacobs, J., *Human Ecology: Research and Applications*, SHE.

CLAES (The Latin American Centre of Social Ecology), 1991, *Membership booklet*, CLAES, Montevideo.

Clancy, K.L., 1990, Agriculture and Human Health, in Edwards, C.E. (ed.), *Sustainable Agricultural Systems*, pp. 655–665.

Clapham, W.B., *Human Ecosystems*, Macmillan, New York.

Cliff, A., 1988, *Atlas of Disease Distributions: Analytic Approaches to Epidemiological Data*, Blackwell, Oxford.

Commission for the Future, 1990, *Our Common Future, The World Commission on Environment and Development*, Oxford University Press, Australia.

Commonwealth Department of Human Services and Health, 1994, *Better Health Outcomes for Australians*, AGPS.

Cox, B.D., Blaxter, M., Buckle, A., Fenner, N., Golding, J., Fore, M., Huppert, F., Nickson, J., Roth, M., Stark, J., Wadsworth, M. and Whichelow, J., 1987, *The Health and Lifestyle Survey*, Health Promotion Research Trust, London.

Crichton, A., 1990, *Slowly Taking Control?*, Allen & Unwin, Sydney.

Crawford, R., 1993, A Cultural Account of Health: Control, Release and the Social Body, in: Beattie, A. *et al.* (eds), *Health and Wellbeing: a Reader*, Macmillan, London.

Dansereau, H.K., 1961, On Implications of Modern Highways for Community Ecology, in: Theodorson, G.A. (ed.), *Studies in Human Ecology*, pp. 175–187.

Dansereau, P., 1994, *An Ecological Analysis of Living Quarters*, Paper presented at the INTECOL, Manchester.

Dunham, H.W., 1937, The Ecology and the Functional Psychoses in Chicago, *American Sociological Review*, 2, August: 467–79.

Day, L.H. and Rowland, D.T., 1988, *How Many More Australians?* Longman Cheshire, Melbourne.

Daysh, Z., 1988, The Work of the Commonwealth Human Ecology Council, in: Borden, R. and Jacobs, J., *Human Ecology: Steps to the Future*, SHE, pp. 19–32.

Daysh, Z. (ed.), 1989, *Human Ecology, Environmental Education and Sustainable Development*, Commonwealth Human Ecology Council (CHEC), London.

Davis, A. and George, J., 1988, *States of Health: Health and Illness in Australia*, Harper & Row, Sydney.

Devanesen, D., Furber, M., Hampton, D., Honari, M., Kinmonth, N. and Peach, H., 1986, *Health Indicators in the Northern Territory of Australia*, Northern Territory Department of Health, Darwin, Australia.

Divakaran-Brown, C. and Honari, M., 1990, *Aboriginal Health in South Australia*, South Australian Aboriginal Health Organisation, Adelaide.

Divakaran-Brown, C. and Honari, M., 1990, Mortality in the Aborigines of South Australia in: Gray, A. (ed.), *A Matter of Life and Death: Contemporary Aboriginal Mortality*, Aboriginal Studies Press.

Downie, R.S., Fyfe, C. and Tannahill, A., 1992, *Health Promotion: Models and Values*, Oxford Medical Publications.

Eckholm, E., 1977, *The Picture of Health, Environmental Sources of Disease*, Norton, New York.

Edwards, C.E., 1985, *Human Ecology: an interaction between man and his environment* (10th edn), Howard University, Washington DC.

Edwards, C.E., 1986, *Human Ecology Monograph* (10th edn), Howard University, Washington DC.

Egger, G., Spark, R. and Lawson, J., 1990, *Health Promotion Strategies and Methods*, McGraw-Hill, Sydney.

Ehrenrich, J., 1987, *Cultural Crisis of Modern Medicine*, Monthly Review Press, New York.

Eisenburg, L. and Sartorious, N., 1990, Human Ecology, *World Health*, Jan–Feb, 28–9.

Elmer, M.C., 1933, Century old ecological studies in France, *American Journal of Sociology*, 39, July: 63–70.

Elstheest, J., 1987, Children and Their Health, in: Kelly, P.J. and Lewis, J.L., 1987, *Education and Health*, Pergamon, pp. 53–59.

Elsworth, S., 1990, *A Dictionary of the Environment: A Practical Guide to Today's Most Important Environmental Issues*, Paladin Grafton Books, London.

Ferre, Frederick, 1988, Eco-technics and the threat of cost benefit thinking, in: *Human Ecology: Research and Applications*, SHE.

Freund, P. and McGuire, M., 1991, *Health, Illness and the Social Body*, Prentice-Hall.

Freedman, B., 1989, *Environmental Ecology*, Academic Press.

Gabarino, J., 1992, *Children and Families in the Social Environment*, Aldine de Gruyter, New York.

Gadow, S., 1992, Existential Ecology: the Human and Natural World, *Social Science and Medicine*, 35(4): 597–602.

Gardner, H., 1989, *The Politics of Health: The Australian Experience*, Churchill Livingstone, Melbourne.

Gettys, W.E., 1940, Human Ecology and Social Theory, *Social Forces*, 18, May: 469–76.

Glaeser, B., 1988, A Holistic Human Ecology Approach to Sustainable Agricultural Development, *Future*, December, 671–678.

Goldsmith, E., 1988, *The Great U-Turn De-industrialising Society*, Green Books, Devon.

Gordon, A. and Suzuki D., 1990, *It's a Matter of Survival*, Allen & Unwin, Melbourne.

Goudie, A., 1990, *The Human Impact on the Natural Environment*, Blackwell, Oxford.

Goldman, N., 1994, Social Factors and Health, *Proceedings of the National Academy of Science of the USA*, 9(4): 1251–5, 15 Feb.

Goldsmith, E. and Hilliard, W., 1988, *Battle for the Earth: Today's Key Environmental Issues*.

Goldstein, G., Novick, R. and Schaefer, M., 1990, Housing, Health and Wellbeing, *Journal of Sociology and Social Welfare*, 17(1) March: 161–181.

Golley, F.B., 1993, Foreword: General understanding and role of ecology in education, in: Hale, M., *Ecology in Education*, Cambridge University Press.

Gough, M., 1991, Human Health Effects: What the Data Indicate, *Science of the Total Environment*, 104(1–2): 201–209.

Graham, J. and Honari, M., 1991, Human Ecology and Health Advancement, Paper presented at the International Conference on Human Responsibility and Global Change, Goteborg, Sweden; published in *International Journal of Environmental Education and Information*, University of Salford, UK, 1992, Vol. 11, No. 4, pp. 233–258.

Grant, C. and Lapsley, H.M., 1990, *The Australian Health Care System*, University of New South Wales Press, Sydney.

Green, J.L., 1969, *Economic Ecology – Baselines for Urban Development*, University of Georgia Press, Atlanta.

Groombridge, B., 1992, *Global Biodiversity: Status of the Earth – Living Resources*, Chapman & Hall, London.

Haines, A. and Parry, M., 1993, Climate Change And Health *Journal of Royal Society of Medicine*, 86(12): 707–11, December.

Hale, M., 1993, *Ecology in Education*, Cambridge University Press.

Hall, R.H., 1990, *Health and the Global Environment*, Basil Blackwell, Oxford.

Hancock, T., 1993, Health, Human Development and Community Ecosystem: Three Ecological Models, *Health Promotion International*, 8(1), 41–47.

Hansson, L.O. and Jungen, B., 1992, *Human Responsibility and Global Change*, Proceedings of the International Conference in Goteborg, University of Goteborg, Sweden.

Harrison, J.L., 1988, Human Nutrition Ecology, in: *Human Ecology, Research and Applications*, SHE, 266–270.

Hassenteufel, P., 1994, *Health Policy in France: the case of cost containment*, Paper prepared for Workshop on the State and the Health Care System, 17–22 April.

Hawke, R., 1989, *Our Country Our Future: Statement on the Environment*, AGPS, Canberra.

Hawley, A.H., 1944, Ecology and Human Ecology, *Social Forces*, 22, May: 398–405.

Hawley, A.H., 1950, *Human Ecology: a theory of community structure*, Ronald Press, New York.

Hawley, A., 1973, Ecology and Population, *Science*, 179, March, 1196–1199.

Hawley, A.H., 1986, *Human Ecology: a theoretical essay*, University of Chicago Press.

Hayes, L.C. and Rodenbeck, S.E., 1992, Developing a Public Health Assessment, *J Environmental Health*, 55(2): 16–18.

Health Department of NSW, *Health for All – promoting health and preventing disease in New South Wales*, State Health Publications, Sydney.

Health Target and Implementation Committee (Health for All), 1988, *Health For All Australians*, Report to the Australian Health Ministers Advisory Council and the Australian Ministers' Conference, AGPS, Canberra.

Helman, C., 1984, *Culture, Health and Illness*, Wright, London.

Hogg, C., 1991, *Healthy Change*, Socialist Health Association.

Hollingshead, A.B., 1939, Human Ecology, in Park, R. (ed.), Principles of Sociology, Barnes & Noble, New York: 63–168.

Hollingshead, A.B., 1940, Human Ecology and Human Society, *Ecological Monographs*, 10: 354–366.

Honari, M., 1971, Importance du Palmier-Dattier dans la Vie des Habitants de Khur, *Objet et Monde*, Paris, Vol. XI: Spring.

Honari, M., 1972, Water and Irrigation, *Journal of Iranian Ethnology*, Vol. I, Iranian Centre for Anthropology, Tehran.

Honari, M., 1973, Camel Herding in the Desert, *Journal of Iranian Ethnology*, Vol. II, Iranian Centre for Anthropology, Tehran.

Honari, M., 1973, Talavak: A Village in the Alburz Mountains, unpublished manuscript, Archive of the Iranian Centre for Anthropology.

Honari, M., 1973, Khoor: A Town in the Central Desert, unpublished manuscript, Archive of the Iranian Centre for Anthropology.

Honari, M., 1973, *Owsungon, A Collection of Folk-tales*, Iranian Centre for Anthropology, Tehran.

Honari, M., 1975, *Analysis of Names of Places in the Semnan Province*, Geographical Organisation of Iran, Tehran.

Honari, M., 1975, *Do Shah-Rah e Kaviri, Study of Two Ancient Roads in the Central Desert of Iran*, Geographical Organisation of Iran, Tehran.

Honari, M., 1975, *Tazieh dar Khoor, Religious Ceremonies in Khoor*, Iranian Centre for Anthropology, Tehran.

Honari, M., 1977, Growth of Human Settlements around the Desert, in *Proceedings of the Eighth Congress of Iranian Studies*, Iranian National University, Tehran.

Honari, M., 1977–1978, History and Spread of Qanat System of Water Supply in the World (seven articles), *Honar-o-Mardom*, Oct., Nov., Dec., 1977; March, April, May, June, 1978, Tehran.

Honari, M., 1979, *Qanats and Human Ecosystems in Iran*, PhD Dissertation, Centre for Human Ecology, University of Edinburgh, UK.

Honari, M., 1980, *Contemporary Environmental Issues*, Research and Publication Bureau, Department of Education, Tehran.

Honari, M., 1980, *Rural Societies in Iran*, Course Text, Tehran College of Social Works, Tehran.

Honari, M., 1988, Health Indicators and Health Impact Assessment, VII Annual Meeting of the International Association for Impact Assessment, Griffith University, Brisbane, 5–9 July.

Honari, M., 1989, Qanats and Human Ecosystems in Iran, in: Beaumont, P., Bonine, M. and McLachlan, K. (eds), *Qanats, Kariz and Khatara*, Menas Press, London.

Honari, M., 1990, Causes of Aboriginal Mortality in the Mid-1980s, in: Gray, A. (ed.), *A Matter of Life and Death, Contemporary Aboriginal Mortality*, Aboriginal Studies Press, Canberra.

Honari, M., 1991, Health Research, New Direction, Keynote speaker, Shiraz University of Medical Sciences, Shiraz, 13 July.

Honari, M.,1991, Environment and Development, Pre-CHOGM Seminar, 29–31 August, Commonwealth of Learning, Vancouver, Canada.

Honari, M., 1991, Changing World, Changing Environment, Centre for Environmental Studies, Tehran University, Keynote speaker, July.

Honari, M., 1992, Ethics and Health Research, Invited Keynote Address, National Workshop on Health and Medical Research Design, Research Division, Ministry of Health and Medical Education and Tehran University of Health and Medicine Sciences, 24–25 December, Karaj, Iran.

Honari, M., 1993, Advancing Health Ecology: where to from here, in Newman, N. (ed.), *Health Ecology: a nursing perspective*, Proceeding of the First National Nursing and the Environment Conference, Melbourne.

Honari, M., 1993, Human Ecology and Development, in: Daysh, Z. and Hall, D. (eds), *A Dialogue of Cultures for Sustainable Development*, Commonwealth Human Ecology Council, London.

Honari, M., 1993, Health Research and Health Systems Planning, Invited Keynote Address, National Conference on Health and Vital Statistics, Tehran University of Health and Medicine, January, Varamin, Iran.

Honari, M., 1994, Health Ecology, paper presented at 'Health and the Urban Environment: promoting good practice globally', a symposium organised by the UK Health for All Network, Public Health Trust and the British Council, 29–30 June, Manchester.

Honari, M., 1994, Human Ecology and Well-being, papers presented at the symposium on Human Ecology, IV International Congress of Ecology, 25 July, Manchester.

Honari, M., 1995, Health of the Whole is Health for All, *CHEC Journal*, No. 12, November, pp. 8–10.

30 *Morteza Honari*

Honari, M., 1997, *Environmental Impact Assessment: Capacity Building and Institutional Strengthening in Iran*, report and proposal on EIA in Iran, United Nations Development Program.

Honari, M. and Graham, J., Keynote Address, 1993, A Framework for Excellent Practice in Community Occupational Therapy, National Conference of Occupational Therapy, University of Rehabilitation, 22–24 May, Tehran.

Honari, M. and Webb, Y., 1992, Diet, Human Ecosystems and the Quality of Life, Poster presented to the Dieticians Association of Australia, 11th National Conference, Canberra, 7–9 May.

Hook, C.N. and Paolucci, B., 1970, The Family as an Ecosystem, *Journal of Home Economics*, 62: 315–318.

Huxley, J., 1934, Applied Science of the Next Hundred Years, *Life and Letters*, Vol. 11, October: 40.

Hyland, M.E., 1993, The Validation of Health Assessment, *Journal of Health Clinical Epidemiology*, 46(9): 1019–23, September.

Ineichen, B., 1993, *Homes and Health*, Spon, Chapman & Hall, London.

Jackson, B., 1994, *Poverty and the Planet: A Question of Survival*, World Development Movement, Penguin.

Johnson and Heuy, D., 1990, Building Human Ecology into Political Action: a Challenge, in: Borden, R. and Jacobs, J., *Human Ecology: Steps to the Future*, SHE, pp. 24–30.

Jones, Greta, 1986, *Social Hygiene in Twentieth-century Britain*, Croom Helm, London.

Jones, P.S. and Meleis, A.I., 1993, Health in Improvement, *Advances in Nursing Science*, 15(3): 1–14, March.

Kelly, P.J. and Lewis, J.L., 1987, *Education and Health*, Pergamon, Oxford.

Kennedy, P., 1993, *Preparation for the Twenty-first Century*, HarperCollins, London.

Khaliqi, M. and Honari, M., 1975, *Analysis of the Traditions Related to Women*, Third Seminar on Family and Culture, Tehran University and Ministry of Culture and Arts.

Khaliqi, M. and Honari, M., 1974, Analysis of the Traditions Related to Children, In *Farhang o Khanavadeh, Family and Culture*, Proceedings of Seminar on Family and Culture, University of Tehran and Ministry of Culture and Arts, Tehran.

Kickbusch, I., 1989, Approaches to an ecological base for public health, *Health Promotion*, 14(4).

Kilsdonk, A.G., 1983, *Human Ecology: Meaning and Usage*, College of Human Ecology Monograph series 102, East Lansing, College of Human Ecology, Michigan State University.

Krader, Lawrence, 1955, Ecology of Central Asian Pastoralism, *The Southwestern Journal of Anthropology*, 11, Winter: 301–326.

Krieps, R., 1989, *Environment and Health: A Holistic Approach*, Luxembourg Ministry of Environment, Gower, London.

Kormondy, E.J., 1984, *Concepts of Ecology*, Prentice-Hall, NJ.

Labonte, R., 1993, A Holosphere of Healthy and Sustainable Community, *Australian Journal of Public Health*, 17(1): 4–12.

Last, J.M., 1987, *Public Health and Human Ecology*, Appleton & Lange, Sydney.

Lawson, J.S., 1991, *Public Health Australia – An Introduction*, McGraw-Hill, Sydney.

Learmonth, A., 1988, *Disease Ecology: An Introduction*, Blackwell, Oxford.

Lee, K.L, Schwarz, E. and Mak, K.Y., 1992, Improving Oral Health through Understanding of Health, *International Dental Journal*, 43(1) Feb.: 2–8.

Levi, L., 1993, Conditions of Life, Lifestyle and Health in a Highly Developed Country, Seish shinkin zasshi, *Psychiatria et Neurologia Japanica*, 95(3), 259–70.

Levin, Y. and Lindesmith, A., 1937, English Ecology and Criminology of the Past Century, *Journal of Criminal Law and Criminology*, 27, March: 801–815.

Litva, A. and Eyles, J., 1994, Health or Healthy: Why People are not Sick in a Southern Ontarian Town, *Social Science and Medicine*, 39(8), 1083–1091.

Loening, Urlich, 1990, Thoughts on humanity and science, in *Human Ecology: Steps to the Future*, SHE, pp. 45–51.

Lovelock, J.E., 1987, *Gaia: A New Look at Life on Earth*, Oxford University Press, Oxford.

Lovelock, J.E., 1989, *The Ages of Gaia: A Biography of Our Living Earth*, Oxford University Press, Oxford.

Lowry, Stella, 1991, *Housing and Health*, BMJ, London.

Luster, T. and Okagaki, L., 1993, *Parenting: An Ecological Perspective*, Lawrence Erlbaum, New Jersey.

Mattingley, C. and Hampton, K., 1988, *Survival in Our Own Land: 'Aboriginal' experiences in 'South Australia' since 1836*, Aboriginal Literature Development Assistance Association Inc, Adelaide.

MacNeill, J., Winsemiu, P. and Yakushiji, T., 1991, *Beyond Interdependence*, Oxford University Press, Oxford.

Max-Neef, M., 1988, *What's Wrong with the Health System*; Economics, Politics and Health; The Challenge of Future Trends, The Health Issues Centre.

McDonald, M.J. and Pickett, T.A., 1993, *Human as Components of Ecosystems: the Ecology of Subtle Human Effects and Populated Areas*, Springer-Verlag, New York.

McGinnis, J.M., 1994, The Term 'Years of Healthy Life', *American Journal of Public Health*, 84(5): 865–867.

McKenzie, R.D., 1926, The Scope of Human Ecology, *Proceedings of the American Sociological Society*, 20: 141–154.

McMichael, A.J., 1993, *Planetary Overload: Global Environmental Change and the Health of the Human Species*, Cambridge University Press, Cambridge.

Mercier, G., 1993, Health, Illness and Christian faith, *Nurses International*, 9(2): 6–7.

Michaels, J., 1974, On the Relationship Between Human Ecology and Behavioural Social Psychology, *Social Forces*, 52, March: 313–320.

Micklin, M. and Choldin, H.M., 1984, *Sociological Human Ecology; Contemporary Issues and Applications*, Westview Press, Boulder, CO.

Milbrath, L.W., 1991, Envisioning a sustainable society: social learning approach, in Sontag, M.S., *Human Ecology, Strategies for the Future*, SHE.

Moran, M., 1994, Three Faces of the Health Care State, paper prepared for workshop on the State and the Health Care System, 17–22 April.

Morone, J.A., 1994, Why is There no National Health Insurance in the United States? paper prepared for workshop on the State and the Health Care System, 17–22 April.

Nath, B., Hens, L. and Devuyst, D., 1993, *Environmental Management: Instruments for Implementation*, VUB Press, Brussels.

National Aboriginal Health Strategy Working Party, 1989, *A National Aboriginal Health Strategy*, Australia.

National Health and Medical Research Council, 1991, *Ecologically Sustainable Development, The Health Perspective*, Australia.

National Health Strategy, 1992, *Enough to make you sick: How income and environment affect health*, National Health Strategy, Research paper No. 1.

Nayar, S., 1987, Health Education in Rural Areas, in, Kelly, P.J. and Lewis, J.L., *Education and Health*, Pergamon, Oxford, pp. 47–52.

O'Keefe, E., Ottewill, R. and Wall, A., 1992, *Community Health*, Business Education, UK.

O'Neill, M.A., 1994, Entrenched Interests and Exogenous Change: Doctors, the State and the Reform of Publicly Funded Health Care in Canada and the United Kingdom, paper prepared for workshop on the State and the Health Care System, 17–22 April.

Owen, D.F., 1974, *What is Ecology?*, Oxford University Press.

Pagel, H., 1988, *The Dream of Reason*, Simon & Schuster.

Pang, K.C., 1987, Health Education in Schools in Hong Kong, in: Kelly, P.J. and Lewis, J.L. 1987, *Education and Health*, Pergamon, Oxford, pp. 129–11.

Paolucci, B., 1980, Evolution of Human Ecology, *Human Ecology Forum*, 10: 17–21.

Paolucci, B., Hall, O.A. and Axinn, N., 1977, *Family Decision Making: An Ecosystem Approach*, Wiley, New York.

Park, R.E., 1915, The City: Suggestions for Investigation of Human Behaviour in the City Environment, *American Journal of Sociology*.

Park, R.E., 1936, Human Ecology, *American Journal of Sociology*, 42:1–15, repr. in: Young, G. (ed.), 1983, *Origins of Human Ecology*, Hutchinson, Stroudsburg.

Park, R.E., Burgess, E.W. and McKenzie, R.D., 1925, *The City*, Chicago University Press.

Pedler, K., 1991, *The Quest for Gaia: A Book of Changes*, Paladin Grafton Books, London.

Phillips, J.R., 1995, Changing Family Patterns and Health, *Nursing Science Quarterly*, 6(3): 113–4, Fall.

Pratt, J. and Young, G.L., 1990, *Human Ecology: Steps to the Future*, SHE.

Public Health Association of Australia, 1991, *Report of the Ecologically Sustainable Development and Health Consultation Project*, PHA, Canberra.

Quinn, J.A., 1939, The Nature of Human Ecology, *Social Forces*, 18: 161–168.

Quinn, J.A., 1940, Human and Interactional Ecology, *Amer. Sociological Review*, 5: 713–22.

Raffestine, C. and Roderick, L., 1990, An Ecological Perspective on Housing, Health and Wellbeing, *Journal of Sociology and Social Welfare*, 17(1), 143–160.

Ramsay, Maureen, 1994, *Ethical Dilemmas: making choices*, paper prepared for workshop on the State and the Health Care System, 17–22 April.

Ranson, R., 1993, *Health Housing*, Spon, WHO, Chapman & Hall, London.

Robinson, W.S., 1950, Ecological Correlation and the Behaviour of Individuals, *American Sociological Review*, 15, June: 351–357.

Robottom, I., 1993, The role of ecology in education: an Australian perspective, in: Hale, M., *Ecology in Education*, Cambridge University Press.

Rojo, T., 1991, Sociological Contribution to Human Ecology: A European Perspective, in Sontag, M.S., Wright, S. and Young, G., *Human Ecology, Strategies for the Future*, SHE, pp. 84–98.

Saint-Yves, I.M.F. and Honari, M., 1988, Skin Cancer in Australia's Northern Territory 1981–1985, *Journal of the Royal Society of Health*, Vol. 108: April.

Sajiwandani, J., 1993, Exploring the Process of Learning in Human Ecology, *Nurse Education Today*, 13(5) October: 349–61.

Sayce, R.U., 1938, The Ecological Study of Culture, *Scientia*, 63, May: 279–85.

Sax, S., 1990, *Health Care Choices and the Public Purse*, Allen & Unwin, Sydney.

Schell, L.M., Smith, M.T. and Bilsborough, A., 1993, *The Urban Ecology and Health in the Third World*, Cambridge University Press, Society for Human Biology Series.

SHE (Society for Human Ecology), 1991, *Membership booklet*, SHE.

Shrader-Frechette, K., 1990, *Environmental Ethics, Human Health and Sustainable Development: A Background Paper*, WCHE/2/14.

Smith, G., 1990, *Toxic Cities*, New South Wales University Press, Sydney.

Solorzano, A., 1993, A Human Ecological Paradigm of Health, in: Wright, S. *et al.*, *Human Ecology: Crossing Boundaries*, SHE, pp. 287–299.

Sontag, M.S. and Bublz, M.M., 1988, A Human Ecological Perspective for Integration in Home Economics, in: Borden, R.J. and Jacobs, J., *Human Ecology: Research and Applications*, SHE, pp. 117–128.

Sontag, M.S., Wright, S. and Young, G., 1991, *Human Ecology, Strategies for the Future*, SHE.

Stoll, H., 1993, Clarification of the Concept of Health, What Is Health? (Kran akenp Flege-Soin Infir Miers), 86(11): 48–51, Nov.

Straus, D., 1990, Human Ecology: The Speciality of Generalising, in: Pratt, J. and Young, G.L., *Human Ecology: Steps to the Future*, SHE, pp. 13–23.

Suzuki, S., Borden, R. and Hens, L., 1990, *Human Ecology: Coming Age, An International Overview*, INTECOL, Yokahama, Japan/Belgium.

Sword, W., Noesgaard, C. and Majumda, B., 1994, Enhancing Students' Perspective of Health through Non-Traditional Community Experience, *Journal Of Clinical Nursing*, 3(1): 19–24, Jan.

Szalai, A. and Andrews, F.M., 1980, *The Quality of Life: Comparative Studies*, Sage Publications, San Diego.

Their, H., 1987, From Illness to Wellness, in, Kelly, P.J. and Lewis, J.L., *Education and Health*, Pergamon, Oxford, pp. 39–41.

Theodorson, G.A., 1961, *Studies in Human Ecology*, Harper & Row, New York.

Thomson, N. and Honari, M., 1988, Aboriginal Health; A Case Study, in *Australia's Health: the first biennial report by the Australian Institute of Health*, Australian Government Publication Service, Canberra.

Townsend, P., Phillimore, P. and Beattie, A., 1988, *Health and Deprivation: Inequality and the North*, Croom Helm, London.

Turnbull, R.G.H., 1992, *Environmental Health Impact Assessment of Development Projects: a Handbook for Practitioners*, Elsevier Applied Science for WHO, University of Aberdeen.

Turner, B.S., 1987, *Medical Power and Social Knowledge*, Sage, London.

UN, 1954, *International Definition and Measurement of Standard and Levels of Living*.

UN, 1959, *Utilisation of Space in Dwelling*, United Nations Economic Commission for Europe, Geneva.

UNEP, 1989, *Environmental Data Report*, Blackwell.

UNICEF, *The State of the World's Children*, Oxford University Press (various issues).

Walker, B., 1992, Environmental Health in the 21st Century, *Journal of Environmental Health*, 55(3): 36–39.

Warren, H.V., 1989, Geology, trace elements and health, *Social Science and Medicine*, 29(8): 923–926.

Watts, N., 1994, A contextual approach to the analysis of environmentalism, Paper prepared for the European Consortium for Political Research, Madrid 17–23 April 1994.

Wedel, W., Some aspects of human ecology in the central plains, *Amer Anthropologist*, 55, October: 499–514.

Westney, O.E., 1988, Human ecology: concepts and perspective, in: Borden, R.J. and Jacobs, J., 1988, *Human Ecology: Research and Applications*, SHE, pp. 129–137.

WHO, 1992, *Our Planet, Our Health*, Report of the WHO Commission on Health and Environment, Geneva.

WHO, 1993, Life Style and Health, in: Beattie, A. *et al.* (eds), *Health and Wellbeing: a Reader*, Macmillan, London.

WHO, 1961, *Public Health Aspects of Housing*, Technical Report series No. 225.

WHO, 1967, *Appraisal of the Hygienic Quality of Housing and its Environment*, Technical Report series, No. 353.

WHO, 1974, *Use of Epidemiology in Housing Programs and in Planning of Human Settlements*, Technical Report series, No. 544.

WHO, 1987, *Housing and Health: An agenda for action*, Geneva.

WHO, 1991, *Environmental Health in Urban Development*, Technical Report Series, 807.

WHO, 1993, *The Health of Europe*, 49, 1–18.

Wiesner, D., 1992, *Your Health, our World: The impact of environmental degradation on human wellbeing*, Unity Press, Lindfield, Australia.

Wirth, L., 1945, Human Ecology, *American Journal of Sociology*, 50, May: 483–488.

Wolanski, N., 1988, Health: an Ecological Problem, in: Borden, R.J. and Jacobs, J., *Human Ecology: Research and Applications*, SHE, pp. 243–258

Wolanski, N. and Dziewiecki C. (eds), 1991, *Studies in Human Ecology*, Vol. 9, Polish Academy of Sciences Institute of Ecology.

Wright, S.D. and Herrin, D.A., 1988, Conceptual Issues in Understanding Family Dynamics: Studying Family as an Ecosystem or is it the Ecological Study of the Family?, in Borden, R. and Jacobs, J., *Human Ecology, Research and Applications*, SHE.

Wright, S., Dietz, T., Bordon, R., Young, G. and Guagnano, G., 1993, *Human Ecology: Crossing Boundaries*, SHE.

Young, B., 1987, The child-to-child approach: health scouts, in, Kelly, P.J. and Lewis, J.L., *Education and Health*, Pergamon, Oxford, pp. 155–160.

Young, G.L., 1974, Human ecology as an interdisciplinary concept: a critical inquiry, *Advance in Ecological Research*, 8: 1–105.

Young, G.L., 1983, *Origins of Human Ecology*, Stroudsburg, PA, Hutchinson Ross.

Young, G., 1991, Minor Heresies in Human Ecology, in Sontag, M.S., Wright, S. and Young, G., 1991, *Human Ecology, Strategies for the Future*, SHE, 11–23.

Part I

Health in macro ecosystems

2 Good planets are hard to find

Ilona Kickbusch

The concept of an ecological public health

One of the key characteristics of health promotion and the new public health is that it is ecological (Milio, 1987; Martin and McQueen, 1989). The Ottawa Charter for Health Promotion (1986) speaks of a socio-ecological approach to health, and the 'Adelaide Recommendations: healthy public policy' (1988) propose to link the ecological movement and the movement for a new public health. It is necessary, however, to clarify conceptually an ecological public health, to trace the theoretical roots and development of an ecological paradigm in health, and to consider the long-term consequences of an ecological approach to public health.

The concept of an ecological public health has emerged in the past decade in response to a new range of health issues and problems in developed countries. This change can be described as a shift in risk patterns. There are new global ecological risks (such as destruction of the ozone layer and a wide range of environmental hazards and disasters) that pose a risk to health, and health risks are associated with the social, cultural and economic organisations of these societies. These risk patterns tend to be cumulative, have no clear cause and do not allow for simple, straightforward cause–effect interventions. In many cases they tend to be global and finite. Once contracted they can be diagnosed, sometimes alleviated, rarely cured. They generally build up silently and invisibly over time and then emerge as a breakdown in people's bodies and in the social and physical environment. The intervention modes of public health seem ill prepared for this new reality and the risks it poses to the health of populations. This shift has led to a reconsideration of the interdependence of people, their health and their physical and social environments, best illustrated by the mandala of health (Hancock and Perkins, 1985). Building on holistic health approaches developed in the context of the wellness movement, the mandala of health attempts to emphasise the interaction between the mind, body and spirit that constitutes health, but also relates health to the wider concept of an ecosystem that strives for balance. This interaction and interdependence is central to ecological thinking. A new public health approach would therefore not only shift from its present reliance on behavioural epidemiology and surveillance to a more environmental and social approach, but would aim to tackle the risk patterns with an ecological approach.

The Ottawa Charter for Health Promotion, and *Our Common Future*, the report of the World Commission on Environment and Development, also called the Brundtland report (1987) outline a global agenda for change based on a strategy for sustainable development that focuses on health, environment and economy. Together, these documents outline a new public health agenda for the twenty-first century and re-establish the link between public health and social reform. The nature of the challenge of health has changed and the orientation and priorities of public health must therefore change.

C.E.A. Winslow (1923) defined public health as the science and the art of preventing disease, prolonging life and promoting physical (and mental) health and efficiency through organised community efforts for the sanitation of the environment, the control of community infections, and education of the individual in principles of personal hygiene, the organisation of medical and nursing service for the early diagnosis and treatment of disease, and the development of the social machinery which will ensure for every individual in the community a standard of living adequate for the maintenance of health.

This definition of public health and many others need to be transcended.

Societal developments

Industrialised societies have changed rapidly since Winslow's time. Much of this change has been caused by scientific and technological advances – including their negative consequences, economic growth, new forms of global interaction and exploitation, and new patterns of social organisation, political practice and overall lifestyles. This chapter does not analyse the interdependence of these factors and how they work together to produce social change, technological progress, economic development or crisis, as there are many ideologies, academic studies and ad hoc interpretations to choose from. The societies on which this chapter focuses are called developed (although most of them are in decline) or industrialised (although most of them have moved on to become service societies). The social science definitions range a step further, and include post-industrial or even postmodern societies, describing changes in social organisation and value orientation that are usually not part of the debate on growth and development. Even in economics, it is hard to keep up with the new terminology; for example, rapidly developing countries such as Thailand or least developed countries such as Bangladesh.

Most of the definitions of progress, however, are based on gross national product, which is then equated with increased quality of life, and not on such indicators of quality of life as education, health, employment, housing and the quality of the environment. Interestingly, even *Our Common Future* (World Commission on Environment and Development, 1987) classifies health and education as non-economic variables, thereby reinforcing a narrow view of social and economic investment and resources.

Feminist analysis has long drawn attention to this. A system of economic management and accounting and a system of societal values that defines what (and who) is productive and unproductive which is rooted in nineteenth-century Western thought and interpretations of the world, is spreading to all societies (which can no longer be easily distinguished as capitalist or socialist).

Whereas in the nineteenth century economic growth seemed to lead towards a better and richer society and promised a future, people now believe that it is leading to disaster. This nineteenth-century mode of thinking (which has different interpretations in different philosophies and ideologies) includes:

- the belief that humankind dominates and is separate from nature and can exploit it to the limit since natural resources are unlimited;
- the equation of productive work and integration in the official labour market;
- the belief that science and Science guarantee progress;
- the belief that most phenomena can be explained in terms of cause and effect;
- the belief that the freedom of the individual is paramount; and
- the belief that human beings can adequately adjust to the changes caused by growth and progress.

This interpretation of the world does not solve the problems facing humanity, especially in light of new global environmental issues and health problems.

Nevertheless, the argument that economic growth by itself produces health and well-being is widespread and is propagated by international agencies and their economic support programmes. For example, the International Monetary Fund imposes programmes that sacrifice health to achieve economic growth in countries in financial difficulty. This is aggravated by the belief that improved health Science will ensure better health. To determine the validity of the (often combined) arguments that economic growth improves health and that advances in health Science promote health, it is necessary to look back in history.

The societal response to health risk patterns

> The risks that kill you are not necessarily the risks that anger and frighten you. Risk is the sum of hazard and outrage.
>
> (Peter Sandman, Rutgers University)

The infectious diseases of the nineteenth century borne by air, water and food made early death a part of everyday life. Diseases struck silently and rapidly and could rarely be cured. Nevertheless, between 1850 and the early 1900s the prevalence of infectious diseases declined rapidly. Some of the best data on changes in mortality and morbidity over the last century are available in the United Kingdom, which was also where one of the key explanations originated for the secular decline in mortality between 1850 and 1914. The McKeown (1976, 1980) thesis states that advances in medical science did not cause this decline in mortality, but an overall increase in wages and living standards specifically improved nutrition, which led to greater host resistance and overall better health. The new discipline of social medicine was founded at that time based on this thesis.

Meanwhile, an abundance of new research (collated by Szreter, 1988) has put the McKeown thesis in perspective. This reassessment has found four key elements that must inform any new public health strategy.

- Economic growth and increased wages cause an overall change in morbidity and mortality patterns but, by themselves, do not guarantee that the overall health of the population improves. In fact, health differences persist; recent data on inequalities in health from the United Kingdom show clearly that, in many cases, inequity has increased. Rapid growth and urbanisation can cause previously unknown negative health effects, as happened in industrialising England and is happening in large conurbations and in developing countries.
- Health and well-being are improved through the complex interaction of initiatives in various sectors, such as the improvement of housing and working conditions (including the factory and overcrowding Acts), the introduction of compulsory education, the introduction of systemic public health measures, hygiene education by various organisations (including anti-alcohol campaigns by the labour movement), family planning initiatives, and increasing the social rights of specific population groups such as women, workers and children.
- These measures, laws and systems are not obtained without a struggle for reform. The official history of public health portrays the public health movement as a succession of deeds of dedicated professionals instead of putting the improved health of the population in the context of political and social struggles. The new period covered by McKeown's analysis saw the birth of trade unions and new political parties and the spread of the *Communist Manifesto*, as well as the advent family planning advocates and suffragettes. The struggle for reform (or for revolution, in some cases) also meant that certain social and economic conditions were no longer socially acceptable. The social perception of risk had changed and the political response therefore had also to change.
- Scientific medicine as such contributed little to these secular changes; the technologies that mattered were social intervention and the ingenuity of engineers. This is not intended to belittle the contribution of many public health pioneers trained as doctors who played a key role as medical officers of health. Nevertheless, the public health systems and infrastructures which these pioneers so diligently devised (particularly at the local level, with the strong support of local authorities) made most of the difference.

In discussing policy options based on the risk patterns of the 1990s, it must therefore be emphasised that economic growth alone does not guarantee better health (as many laissez-faire adherents would like people to believe); complex systemic measures and legal reform accompanying economic growth are necessary. The health systems introduced a century ago responded to the health challenges and risk patterns of that time – many of which still exist, especially in the developing world. These systems were a major tool in introducing a new historical phase: the industrial society. Public health measures were socially perceived to be essential for the growth of industrial society. They symbolised this society's ideological promise of increased welfare for all through the unprecedented growth of the wealth of nations.

The link between social change, pressure for social reform (of many political colours) and public health has been lost. Many of the problems facing the old public health (such as diseases linked to poor living conditions) have been replaced by straightforward surveillance and lifestyle related diseases attributed to behaviour. Public health has gradually abandoned its holistic approach and moved into a phase of medical dominance, focusing on behavioural epidemiology, preventive medicine and health education. It has individualised cultural patterns by concentrating on disease categories and principles that explain how risk factors caused diseases (for example, heart disease and high blood pressure require reduced fat consumption and other changes in health behaviour). The practice of public health does not yet correspond to the overall changes in risk patterns. Only recently has public health started to incorporate a social model of health which makes healthier choices easier and improves the social climate for health. The future of health depends on the approaches public health systems use in the next 50 years to deal systematically with the issues that confront societies globally.

A new public health agenda

Over the past 150 years public health has shifted from a holistic approach (as reflected in the sanitary idea) to approaches that increasingly tend to be based on the individual rather than on organised community effort and on a social mechanism that ensures to every individual a standard of living adequate to maintain health.

The destruction of the link between public health and social reform and an overall vision of society has left public health weak and vulnerable. The public health system in the nineteenth century was one of the most powerful tools to promote the development of industrial society without too great a human loss, but it has lost this leading role in present developments. Public health is neither at the centre of present health systems, which are dominated by medical techniques and cures rather than community based health efforts, nor is it yet adequately prepared to lead in solving the new health concerns of the community and the global ecological health challenges of the future.

According to its constitution, the World Health Organisation (1986) is the world's leading public health authority. In the process of its development, however, it has succumbed to the medicalisation of public health. At the end of the 1970s the World Health Organisation was in great danger of becoming the world medical organisation, but with great diligence and vision it presented the world with the goal of *Health for All by the Year 2000* (WHO, 1981) and with a strategy to achieve this goal: primary health care. This was a rediscovery of the two basic principles of public health:

- the need to improve living conditions (providing the prerequisites for health, such as housing, water, income, food and education) and
- the need to build a systemic public health infrastructure, to ensure an organised community effort towards health.

Two further elements were added that are critically important to the new public health: health was considered to be a social goal of government, and a global and not just a national challenge. Health has to be achieved on a world wide scale based on joint global commitments.

The strategy of emphasising primary health care was far too often interpreted as a way to provide basic medical care, rather than focusing societal effort on implementing systemic measures that promote health and prevent disease. To reinforce the statement that achieving health for all means to ensure to every individual a socially and economically productive life and therefore to ensure a political commitment to health, WHO undertook a further initiative to develop a new understanding of public health. The Ottawa Charter for Health Promotion (1986) which was adopted at the first International Conference on Health Promotion in November 1986 summarises these efforts. The Ottawa Charter outlines the conceptual starting points, basic principle and action areas of the new public health.

The Ottawa Charter expresses the need for a new view of health; for the first time, a WHO document included a stable ecosystem and sustainable resources as prerequisites for health. It states that caring, holism and ecology are essential issues in developing strategies for health promotion, and calls on health professionals and decision makers to recognise health and its maintenance as a major social investment and challenge, and to address the overall ecological issue of our ways of living.

The Ottawa Charter has helped to redefine health agendas around the world, and it is a watershed in the development of health promotion and public health (Green and Raeburn, 1988). It is definitely the first document to outline an agenda for the new public health by placing it firmly in the context of new ecological thinking. This is, then, the potential impetus that could impel public health to become part of the vanguard of societal development and to provide tools that help to develop a more sustainable society.

The similarities in priorities of strategies in the Ottawa Charter and the Brundtland report are striking, but they are not coincidental. Each redefines the issues at stake in terms of human and ecological resources and development, and expresses a moral obligation to other living beings and to future generations. Each advocates that investment in health and the environment be based on new priorities and patterns of policy making. Each expresses the need for new infrastructures and legal systems.

WHO has recently reinforced this approach in preparing its contribution to the international efforts towards sustainable development. A World Health Assembly resolution (WHO, 1989) stressed that achieving health for all requires the sustainable use of the world's resources and sustainable social and economic development.

From the sewerage principle to an ecological principle

This is the proper agenda for the new public health. Public health needs to rise above petty squabbles over specialised fields of intervention to a generalist and policy based concern for the health of populations, which can no longer be separated from the social mechanisms that produce risks to health. These mechanisms now do so to a far greater extent than could be imagined when nineteenth-century public health systems were established.

A good illustration of these issues is the sewerage principle. Although the sewerage system was a brilliant, successful and innovative Science in public health, it was a solution adequate to an immediate problem. It was a systemic approach that attempted to cover an entire population rather than special risk groups, but over time it had one crucial drawback: nobody really thought that there could be limits to the supply of water or the capacity of rivers and oceans to cope with all the debris. Nobody had envisaged the sheer amount and toxicity of the debris and the cost of managing it over time. Industry, politicians and citizens think in terms of the sewerage principle. In a simplistic form, this metaphor is as follows:

Regardless of the amount or type of negative side effects produced by economic growth, an invisible hand and a working system are there to receive it and get rid of it. Both citizens and industry flush it down and expect it to disappear. This service is expected to work at minimal cost and with no stench and discomfort. Both governments and nature are expected to have an ever-expanding capacity to ameliorate the side effects of production and economic growth. If something goes wrong, systems (such as medical care) are expected to fix the problem or compensate for the damage done (insurance schemes). The prevention implemented copes with the debris and thus averts the worst short term consequences that could arise. Serious prevention is rarely attempted.

Over the past decade, however, the messages warning that the sewerage principle is breaking down have increasingly been heeded. People are finding that they are bathing in their own debris. This is a critical difference between the old and the new public health: there is no 'away' in which to throw things any more. The problems humanity faces have no precedent or analogue in public health history: destruction of the ozone layer, harmful chemicals in groceries, pollution of ground water by pesticides, genetic engineering, large scale depletion of natural resources and others. These problems have accumulated quietly while some people were joking about the warnings of the Club of Rome (Mesarovic and Pestel, 1974) or the concerns of the greens. The world can no longer be conveniently divided into developed and developing as these problems are global and threaten even the people that have produced them. Simple causation and compensation principles do not apply any more, and many of the problems and conditions cannot be cured. Science has only limited answers, and it is part of the problem.

In addition to these environmental issues, public health is increasingly being confronted by:

- disease patterns linked to social inequities and lifestyles in industrialised societies;
- health problems that are social rather than medical in nature;
- health problems that tend to be cumulative, long term and not amenable to curative measures;
- an increasingly ageing population;
- health care systems that do not respond adequately;
- systems of health care finance that are outmoded and inadequate; and
- a general public that is changing its social perception of health risks and is expressing new expectations.

It is becoming increasingly clear – even to the general public – that this toll of disease and illness is just as preventable as infectious diseases were earlier, if the political and social priorities are changed accordingly. People expect to live long and to be healthy and independent in their old age, but not all are changing their health behaviour to support the focus on wellness and quality of life. The challenge for public health and public policy is to outline the major social, economic and political investments into an organised community effort, and to make proposals for social mechanisms that will ensure the promotion and maintenance of health for all by the year 2000 (Terris, 1987). This means redefining public policy, the common good and individual responsibility (Milio, 1983).

A new definition of public health that would take us beyond Winslow without losing the key issues which health addresses (the health of population, organised community effort and the need for social machinery) could be approached by using the Ottawa Charter for Health Promotion (1986) as a starting point. It could read as follows:

- Public health is the science and art of promoting health. It does so based on the understanding that health is a process engaging social, mental, spiritual and physical well-being. Public health acts on the knowledge that health is a fundamental resource to the individual, to the community and to society as a whole and must be supported by investing soundly in living conditions that create, maintain and protect health.
- Public health has an ecological perspective, is multi-sectoral in scope and uses collaborative strategies. It aims to improve the health of communities through an organised effort based on:

  - advocacy for healthy public policies and supportive environments;
  - enabling communities and individuals to achieve their full health potential; and
  - mediating between differing interests in society to benefit health.

- Public health in infrastructures needs to reflect that it is an interdisciplinary pursuit with a global commitment to equity, public participation, sustainable development and freedom from war.

Theoretical base of an ecological public health

Drawing out the implication of an ecological approach to public health requires tracking the theoretical base and epistemology of ecological thinking. There are two ways to do this.

First, one can trace schools of thought in various disciplines that have contributed to or use ecological thinking and have aimed to explain and understand the interactions between humans and their environments (Catalano, 1979). The two extreme disciplines have been biology and sociology, each of which is wary of the other's subject matter, as reflected in the heated discussions about sociobiology. The social and policy sciences, however, have recently moved on the Occupational Therapy debate a new environmental paradigm that applies ecological thinking to the social and political realm (Buttel, 1986; Dwivedi, 1986; *International Social Science Journal* 1986).

Second, a wide range of literature on health has contributed towards an ecological model of health, ranging from miasma theory (Rosen, 1958) to Lalonde's health field concept (Lalonde, 1974; Raeburn and Rootman, 1988) to models of social health (Cassel, 1976; Kickbusch 1986; Marmot and Morris, 1984).

Bateson's work (1975b, 1979) links these two theoretical strands by providing the idea of pattern. Health would then not be defined in terms of host and agent, person and environment or cause and effect, but as the pattern that connects. An ecological theory of public health must be based on this idea.

Developing a theoretical base for the new public health is challenging and complex. It goes beyond how people and environments fit together and beyond simple ideas of adaptation and balance. It does not just mean shifting attention to the effect of the new threats to health posed by the physical and social environment. It implies another way of interpreting health and the systems that create it. The new public health uses a different set of theoretical premises and methods that are in their beginning stages. Arguments for a more holistic and ecological public health are therefore often supported quite haphazardly. Quotes range from the ancient Greeks to the Australian aboriginals since they seem to express a broader idea of health that concurs with recent rediscoveries. The actions of the pioneers of public health are presented out of context, an example being the famous statement of 'politics being medicine at large' by Virchow. As serious, broad academic studies are lacking, arguments are often constructed with little scientific depth.

The origin of an ecological theory of public health includes disciplines other than health and medical sciences. The goal is to clarify the implications of a new concept of health and of a new concept of public. This dual challenge requires interdisciplinary work, ranging from the biological to the political sciences.

Fields other than medicine that contribute to ecology include biology, anthropology, the social and political sciences, history, history of science, philosophy and literature. Analysing the many disciplines that have contributed to ecology as 'a branch of science concerned with the interrelationship of organisms and their environments' (*Webster's Third New International Dictionary*, 1976) also helps to assess whether these disciplines are experiencing a broader paradigm shift in the direction of explaining the world ecologically. Examining the literature on health and ecology can help to identify the exemplars (Kuhn, 1970) of a theory of public health. Comparing these two lines of enquiry can then help to clarify continuities and discontinuities in the social and scientific understanding of the interaction between humans and their environments and the effect of this interaction on health. For example, is the discourse based on relationships of cause and effect or is it based on considering 'the totality or pattern of relations between organisms and their environment' (the second definition of ecology according to Webster (1976)? Chaos theory (Gleick, 1987) is the most important development of this thesis.

Brewer (1979) has attempted to analyse the policy sciences, ecology and public health by discussing the common characteristics of policy sciences and ecology and relating them to public health. Strategies emerge that outline the challenges to an ecological public health:

- encountering prevailing norms of fragmentation and specialisation in disciplines and professions;
- facing cumulative effects whose combined consequences are very gradual and emerge as a crisis only as thresholds are crossed;
- confronting problems that reach beyond and require the integration of specialised bodies of knowledge; and
- using multiple methods, tightly connecting theory and practice, and focusing on contexts and meaning.

Brewer stresses that the anthropocentrism of public health would bring to ecology an additional valuable element by focusing on human health and well-being. Similar discussions can be traced in the social sciences (Buttel, 1986). In particular, Schnaiberg (1980) has developed a model of interpretation and analysis that enables human action and direction to be integrated into an ecological model, in place of the focus on self-regulation and adaptation that is usually applied in ecological anthropology. The US National Academy of Science (1987) has a very interesting approach that includes human skills and value possibilities as part of the reciprocal influences between humans and physical environments. The social sciences lack a tradition, however, for outlining the processes by which the social structure and human action positively or negatively affect the physical environment.

Nevertheless, a wide range of research on social health, social integration, social support and belonging could explain how breakdown in the ecology of human interaction seriously affects health. (A WHO publication on the new social epidemiology is being prepared (Badura and Kickbusch, in press)). Antonovski's (1979) work on a science of health linked to feelings of belonging and social integration (salutogenis) needs to be given much greater attention than it has received so far. In addition, the word lifestyle needs to be reassessed; the original use of this word in social science was based on the context and meaning of human actions, not on functionalist premises of human behaviour. An ecological theory of health must seriously pay attention to the social ecology of humans and their social development, social relationships, culture, emotions and dreams. Conviviality can be reintroduced (as put forward by Ivan Illich (1973) several years ago) as a measure of social health and well-being.

Finally, the oldest public health problem – inequity in access to health – remains a prominent issue for the new public health, although new forms of social inequity in health have emerged: the rise of single parenting, the feminisation of poverty, and the problems of very old people and homeless young people. These forms of inequity indicate the tremendous social and human resources wasted by societies, not only nationally but even more drastically on a global scale.

Based on the literature on health, much needs to be done to overcome the atheoretical and ahistorical stance of public health. Developing a new perspective requires a realistic view of history. Public health has been less attractive to scholars than the rise and power of the medical profession, even when scholars put forth their criticism of medicalisation. It is necessary to determine the social theories, political ideologies, scientific schools and discoveries that influence public health pioneers through the various stages of public health which have led to its present narrow approach. A political history of public health still needs to be written and much comparative work needs to be done.

For example, it would be very interesting to trace the various influences that have forged the key ideas of the new public health, including the critique of medicine, the wellness movement, mutual and self-help, the women's health movement, the crisis of escalating medical costs, holistic medicine, Eastern philosophies, the demand for social justice, new risk patterns and inadequate political responses. Another example would be to outline the options for public health, depending on whether a functionalist or ecological framework is used. If health is viewed in terms of the concept of humans and machines of nineteenth-century science, working with cause and effect in relation to forces and impact, then public health logically builds its intervention on such premises, using a functional risk factor approach (Terris, 1987). Long term, diffuse and cumulative effects of human activity on health require another theoretical base and other intervention approaches. The models that have been proposed particularly follow the ecology of human development developed by Bronfenbrenner (1979) for child development. Chamberlin (1984) used this model to construct an ecological model of child health services (outlined later).

The pattern that connects

An even more fundamental consideration should be taken into account. In *Steps to an ecology of mind*, Bateson (1975a) warns against 'a mass of quasi-theoretical speculations unconnected with any core of fundamental knowledge'. He argues that any investigation works with two types of knowledge: observations and fundamentals. As Kuhn (1970) has outlined, a paradigm shift comes about when the observations and the fundamentals are mismatched, so that the fundamentals cannot explain the world any more, and propose instead the pattern that connects.

A key step in developing an ecological theory of health is understanding as a pattern or relations rather than as a quantitative outcome. A pattern, however, is not fixed but is (Bateson, 1975a) 'primarily a dance of interacting parts and only secondarily pegged down by various sorts of physical limits and by those limits which organisms characteristically impose'. This makes it possible to understand and analyse health as a process, as proposed by the Ottawa Charter, and to analyse interacting parts in terms of context, meaning and relationships. Bateson (1975b) has outlined 'an abstract idea of what we might mean by ecological health'. He defines human civilisation as:

> A single system of environment combined with high human civilisation in which the flexibility of the civilisation shall match that of the environment, to create an ongoing complex system, open ended for slow change of even (hard programmed) characteristics.

Bateson then explains some of these terms in more detail and his definition of high civilisation is similar to many of the recommendations on sustainable development. His concept of flexibility is highly relevant to the theory and action of the new public health.

Bateson (ibid.) broadens the idea of adaptability from passive adaptation to more active adaptation that involves upper and lower levels of tolerance, beyond which discomfort, pathology or death occur. Under stress, a variable gets close to its lower or upper level of tolerance and begins to lose its flexibility. In doing so, it influences the other variables and the lack of flexibility spreads throughout the system. An obvious example is that an overpopulated society tries to make overpopulation a more comfortable process, which gradually leads to more fundamental ecological pathology. The propensity of systems to undermine their own flexibility has serious political consequences. Social flexibility is a resource as precious as oil or titanium and must be budgeted in appropriate ways, to be used for needed change. Thus, health can only be maintained if it remains part of a process geared towards sustainability.

Many theories should be challenged with these fundamentals. Bateson outlines categories that can explain both physical and social processes within a pattern: flexibility, diversity, context, meaning, levels of tolerance, and form. A new dialogue could then emerge between the natural and the social sciences in outlining an ecological theory of health.

Strategies and approaches

An ecological approach to public health needs to develop strategies that correspond to the new risk patterns. These strategies need to go beyond the tinkering that occurs within the present fragmented systems and approaches and need to find new approaches to local, national and global policies. Two key reports outline a general course of action: *Our Common Future* (World Commission on Environment and Development (1987) and the *Global Strategy for Health for All by the Year 2000* (WHO, 1981). Both reports are strongly oriented towards policy, and call for closer scrutiny of government actions in relation to environment and health. Proposals for action have been developed based on healthy public policy and sustainable development. Their agendas are remarkably similar, which may indicate a broader paradigmatic shift in understanding humans and the world and the values and principles that guide governance. These two reports propose to integrate ecological considerations into political and administrative decision making by using sustainability as a guiding principle. A World Health Assembly resolution (WHO, 1989) stresses that health and sustainable development are not only interdependent but reciprocal. Both the environment and health are seen as social resources, as common property that society has an overall responsibility to protect. In an ecological approach to policy, health is part of the ecological wealth of a society. It becomes one of society's key human resources and thus one of the key indicators of sustainable development. The development of health itself must therefore be sustainable.

This leads to new proposals for social investment and political accountability and to the need for new institutional and legal frameworks. Political ecology could provide an analytical approach to examine, analyse and inform government actions and responses to the new risk patterns. Many of these responses need to be based on global cooperation and will, and simultaneously greatly influence people's daily lives. If equity, conviviality, sustainable development and global responsibility are the guiding principles of an ecological public health, then both governments and individuals face hard choices.

Our Common Future (World Commission on Environment and Development 1987) states that: 'The common theme throughout this strategy for sustainable development is the need to integrate economic and ecological considerations into decision making'. This need to integrate is also the key theme of strategies for healthy public policy, as outlined in the Ottawa Charter and the Adelaide Recommendations (1988): 'Healthy public policy is characterised by an explicit concern for health and equity in all areas of policy and by an accountability for health impact'.

Comparing the proposals of the World Commission on Environment and Development and the recommendations of the Adelaide Conference, healthy public policy and strategies for sustainable development have common characteristics, as they both:

• integrate ecological and health considerations into political decision making;
• emphasise accountability for side effects and the impact of decisions;
• promote intersectorality and integrated strategies for action;
• are committed to equity;
• recognise the need for new legal and institutional reforms;
• promote community knowledge, involvement and support;
• advocate investment in the future and take responsibility for future generations; and
• encourage global concern.

Both reports focus on the serious institutional gaps that face societies responding to new ecological problems, and both recognise the need to invent new systems as an institutional challenge of the 1990s. The key themes of both reports are social equity, social investment and social innovation in health and environment, which are guiding principles for sustainable strategies for health and environment.

Consequences of a new public health strategy

An ecological approach moves health from a matter of individual lifestyle and choice to a broad issue for the community. It starts with a basic question. Where is health created? The ecological answer is that health is created where people live, love, work and play. People create health by interacting with each other and with their physical environments. A public health strategy should thus begin with the settings of everyday life in which health is created (rather than with disease categories) and strengthen the health potential of these settings. This leads to identifying patterns that constitute health and developing strategies that strengthen such patterns throughout the process of human development. The Ottawa Charter for Health Promotion (1986) suggests that such patterns are strengthened by a public health strategy that promotes:

- an awareness of public policies and their effects on health;
- social and physical environments that support health;
- personal skills development;
- community involvement; and
- public health services that are responsive and oriented towards health.

For example, a strategy to improve the health of schoolchildren would aim to put health into practice as part of the overall school setting and activities and not just as an activity called health education. This could include:

- teaching personal skills and autonomy;
- promoting a positive body image;
- creating a positive social and physical environment for learning;
- involving the community in school activities (using the school facilities for community events, for evening classes on health and environmental issues, and as an information centre on pollution);
- providing healthy school meals and facilities;
- supporting positive interaction between children and parents; and
- ensuring ecological disposal of school refuse.

The local school would literally become one of the health centres of the town. Industrialised societies are obsessed with creating health centres staffed with medical and paramedical personnel rather than looking for social entities that could be centres of health and supporting them in such a role. Chamberlin (1984) has outlined such a strategy:

Child health and developmental outcomes are related to parent functioning, which is influenced by both formal (health and human service providers) and informal (family and friends) community support systems. These are in turn influenced by both cultural values and the policies of local, state and national governments. With this approach the focus is on the community as a whole and the relationships between families and their current environments.

The focus of this strategy is to strengthen the community resources that improve the functioning of all families and children, focusing on the total population instead of risk groups.

Chamberlin (1984) proposes five components at community level instead of the highly fragmented and excessively professionalised approaches currently practised:

- a community council to establish priorities and coordinate services;
- a community-wide health education programme;
- the availability of basic support services for parents;
- a consumer advocacy organisation; and
- a reliable assessment system.

Examples of such integrated approaches to child health can be found in the Nordic countries, and the child health statistics show the merit of such an approach. The Victoria Community Health Councils in Australia (Milio 1988) are a move in this direction. Chamberlin (1984) argues that environmental sanitation is the best model of a systemic approach. It would be ineffective if it targeted only high risk groups or relied on each individual boiling water before using it. Many community health projects now use these ecological approaches at the level of social action, but most are isolated examples rather than part of an integrated system of ecological community care. This returns to the sustainability of the development of health systems.

Changing expectations and social perceptions

What are the chances of developing integrated policy approaches for the new public health on a large scale? The political pressure for change has increased recently as the social perception of health and environment problems, and the health-related behaviours and the awareness of health of certain strata of society have changed. For example,

- Environmental destruction is increasingly seen as an acute new social problem.
- Overall consciousness of health has increased among the middle strata of society, although it is still mainly individually oriented (my health rather than community health).
- More and more people are gaining experience in self help and mutual aid groups.

- Middle-class consumers are influencing the range of healthy products and their presentation (such as labelling) and producers must respond to substantial shifts in consumer behaviour.
- Some harmful and hazardous products are being phased out slowly under the pressure of public policy and public opinion (such as tobacco, at least in some developing countries).
- Certain health damaging behaviours are becoming less acceptable and less socially desirable.
- The media show great interest in revealing new scandals related to health and the environment.
- Most importantly, the new global environmental hazards have drawn people's attention to the limited choice for health they can make as individuals.

On the whole, public interest in preventive measures is growing and the public increasingly expects governments to take responsibility for health and environmental hazards. Data from a July 1988 opinion poll in Canada showed that 77 per cent of Canadians said that they would pay more for a product if it were labelled environmentally safe, and that 56 per cent would be willing to pay two cents more for milk or gasoline to help improve the environment.

Another Canadian poll in autumn 1987 had interesting political implications: 92 per cent of those surveyed said that corporate executives should be held personally responsible if their company repeatedly pollutes at unsafe levels; 78 per cent were willing to pay higher taxes or prices to improve environmental protection; and 87 per cent were upset about the lack of action taken to protect the environment.

Political ecology

People increasingly perceive health as a social right (Beck, 1986). This trend is considered to be different from the demands for more services raised in the 1960s and 1970s. Citizens are gradually becoming aware that structural measures and public policies are required to ensure increased health and to reduce hazards such as chemical residues in food, food additives, radiation and pollution. People who have developed a health consciousness as ecologically responsible consumers find it particularly difficult to accept political non-decisions and inertia, as they can see that such responsible individual behaviour is impossible because they cannot control the products for sale, the quality of the air they breathe or the sand in their children's playground. It is also essential to ensure the credibility of the behavioural change models proposed by health education. People can improve their health themselves, but this needs to be reinforced and supported by more comprehensive systemic measures. Healthier choices are rarely easier to make; they are usually more expensive and are only available to a minority of the population, since healthier choices are heavily based on access to information and financial resources. Healthier choices are often still a minority behaviour within cultures that do not make health promotion a priority. Political legitimacy is threatened, as institutions and administrations can no longer fulfil the promise to deliver safety and freedom from harm and to protect the common welfare. This is further exacerbated by an increasing distrust of the medical system as the medical industry has increased its diagnostic capacity but cannot cure many of the most prevalent diseases, including cancer, diseases of the joints, chronic pain and AIDS.

Systems of government are therefore challenged to integrate the increasing concern about the environment and health into their policy proposals and their day-to-day politics. Except for some symbolic measures, however, governments do not seem to know how to respond to the combined pressures arising from the technical, legal, social, political and economic dimensions of the problems facing an angry and confused public and press on the one hand, and established power brokers, interest groups and industry on the other. In such a context, sustainable development and healthy public policy are frequently criticised as being abstract ideas that are impossible to put into practice. They seem idealistic only because there is no experience of putting them into practice and little political will to do so. Unecological and fragmented systems of government cannot see the forest even as the trees are being felled. Many public policies that are taken for granted in industrialised societies were initially considered unfeasible. Imagine the amount of pressure, negotiating and coalition-building it took to cover London with a sewer system.

A concept termed political ecology by Graham Beakhurst (1979) has recently emerged within political science. It aims to highlight the political dimensions of environmental and ecological concern, to discern the forms in which power and authority are exercised in dealing with the new ecological issues, ranging from local to international policies and encompassing action by governments and by non-governmental organisations. As the Brundtland report (World Commission on Environment and Development, 1987) points out, the existing political and economic structures will not ensure human survival; political values and processes need to be transformed to meet the requirements of sustainable development.

Dwivedi (1986) has outlined the plodding development in environmental policies in industrialised countries. Governments in the 1960s and 1970s responded to environmental problems incrementally, using existing agencies to administer solutions. In the early 1970s governments established new administrative units such as environmental protection agencies or ministries for the environment. (In many countries the environmental responsibilities of public health departments were correspondingly reduced). These new units gave environmental issues a new profile and developed new mechanisms such as environmental impact assessments, but their effectiveness continues to be hampered by the economic priorities of governments, despite recent attempts to cooperate globally. Most of these developments were catalysed by increasing public pressure for action. In some countries new political parties with an environmental agenda entered parliament, since the established political parties were viewed as ineffective.

A new form of decision-making is therefore needed that integrates action and public accountability and that re-establishes political credibility. Science cannot provide the answer, as it can provide more uncertainty. For example, it took great scientific ingenuity to discover the hole in the ozone layer in the stratosphere, but there is still no definitive proof of the harmful health effects of increased destruction of this ozone. The preliminary indications include increased rates of skin cancer, chronic changes in lung functioning and suppression of the immune system. Political decision-makers will therefore be increasingly forced to use political and social criteria to assess risks instead of relying on definitive scientific proof. Trevor Hancock (1989) predicted that governments will be forced to respond to the consequences for health of unsustainable agricultural policies, unsustainable energy use, chemical and radioactive contamination, resource depletion and further urbanisation.

These problems and the common global challenges that need to be tackled according to *Our Common Future* present a formidable challenge. The idea of healthy public policy is therefore crucial because many ecological issues have finally become broad public issues as a result of the effects they have on people's health: the effects on their everyday lives, on their children and on their hopes and fears. The increased concern for health caused by the medicalisation of society and the concept of risk factors has turned a somersault. It has produced a public that increasingly demands health and does not accept a lack of preventive action in an area that is clearly governmental responsibility. People used to debate whether a strategy to combat smoking should begin with the individual smoker's freedom to smoke or with the production, pricing, advertising and taxation of tobacco. The tobacco industry ruthlessly exploited this ambivalence by warning that the values of democracy were at stake if tobacco advertising were banned. These issues appear reasonably straightforward compared with the present issues in focus, as individuals can hardly control the harmful substances used to produce the roast and vegetables (for soya steak) for Sunday dinner or at what point of the production process and under whose responsibility toxic chemicals are introduced into the food chain. This raises new issues of legal and political responsibility for harm and new ways to regulate compensation that cannot be explored here (Reich, 1988). Preliminary indications can be seen in the legal measures taken against tobacco companies and against employers who do not introduce non-smoking work areas and subject employees to the risk of passive smoking.

The social perception of risk is changing and is influencing people's political response. Demands for action at international, national, state and local levels will emerge, as will proposals for individual strategies and consumer movements to support sustainability. Returning to the McKeown thesis, the health of populations was changed through economic growth and a wider distribution of wealth, pressure for social change and reform, concerted action in many sectors and the establishment of a population-wide system of prevention. Are these factors sufficient to catalyse change in the health of populations at present? If not, how do they need to be adjusted, for example, changing the distribution of wealth globally rather than nationally? To contribute to a more sustainable society, public health must:

- develop proposals for legal and institutional reform that strengthen the promotion and protection of health;
- ensure that the potential effects on health of the new environmental risks are seriously considered in policy decisions at all levels;
- ensure that assessments of effects on health and the environment become part of governmental planning and accountability;
- ensure that the public is fully informed of risks to their health;
- give priority to reducing inequities in health; and
- open the debate on the sustainability of health development itself.

To accomplish this, public health must develop methods and mechanisms that are accountable to the public and that support integrative strategies. As a guideline for institutional reform and developing strategy, the ecology of systems and increased social flexibility and diversity must be considered a societal resource for survival. Many professionals oppose such developments, as many portfolios are based on restricted ways of solving problems.

A broad social debate should be opened on issues such as:

- Why not have a minister for health who is really in charge of health, has an appropriate budget to do so and must be heard at cabinet level?
- Why not have a ministry of health and the environment and a ministry of social and medical care?
- Why not consider more carefully how a public good such as health is produced and ensure accountability when health is destroyed?
- Why not accept that the skills needed to promote health are totally different from those needed to cure ill people?
- Why not discuss the balance between individual, social and political responsibility for maintaining health, to create a new culture of health and social responsibility?
- Why not introduce an accounting system within government that makes visible the positive and negative effects of other sectors on health and links it to their budgeting?
- Why not lift the taboo on the appropriation of public goods (and money) for professional and private use and reassess the public ownership of the commons?

The list of questions could be expanded. As the Brundtland report says, many public health issues reflect the fact that people need to renegotiate what kind of society they want as they enter the twenty-first century. Is there any consistency in various societal goals? An intense political and social debate on what and who produces health and what people are willing to pay for it would be a positive step.

- Why should health promotion be organised using the principles employed to solve problems in the nineteenth century?

- Why is it acceptable that a middle-class man in Australia lives an average of seven years longer than his poorer counterpart and has less disability and pain all his life?
- Why is it acceptable that men take no responsibility for their children and then proclaim that the family is breaking down?

Public health has some hard choices to make. Bateson (1975b) says that ecologists face the paradox that, to preserve flexibility (for example, preserving irreplaceable natural resources), 'their recommendations must become tyrannical'.

This raises the most complex social and political issue of all. Are there new ways to define, determine and rank the individual and the common good? This is the ultimate question posed by political ecology. Andrei Sakharov stated that, for the former USSR, 'for the individual . . . the question of collective rights is more pressing than that of personal ones' (interview with *Le Figaro*, 1989). The Brundtland report clearly puts the needs of future generations above the economic expansion of present societies: the health-for-all strategy requires a transfer of societal resources towards greater equity, to ensure access to the prerequisites for health and basic health care needs. The report of the Worldwatch Institute (1989) indicates the need to reduce family size worldwide and emphasises that: 'Any meaningful effort to slow population growth will depend on heavy investment in the provision of family planning services, improvements in education and health (particularly for women, one might add) and financial incentives that encourage smaller families'. These proposals conflict with the strategies demanded by such international financial institutions as the World Bank and the International Monetary Fund, which blindly expect countries to reduce their investments in education, welfare and health (which, of course, these institutions call expenditures) to promote growth in gross national product.

It is painful to see the McKeown thesis reduced to the incorrect formula that increased gross national product is equivalent to improving the health of populations. An ecological perspective views the issue exactly the other way around. Health is part of the ecological wealth of a society, one of its key resources, and one of the key indicators for sustainable development. The pattern of health is sustained by the relationships between conviviality, equity and ecology, or as the WHO Constitution stated with great insight in 1948 (WHO, 1988), physical, mental and social well-being.

As health – social, physical, mental and spiritual well-being – is the outcome of a societal pattern, improved health in a society provides information on the general quality of life (context) and the overall values (meaning) of the society. Health describes the interaction between humans and their environment and indicates their specifically human skills and ingenuity, and potential for innovation and caring (flexibility). Finally, health has to be assessed at a global level. National goals are not enough and need to be supported by new types of international cooperation, monitoring and accountability. As Bateson (1975b) says: 'We are not outside the ecology for which we plan – we are always and inevitably a part of it'.

The strategies and mechanisms of the new public health are in their infancy, and must mature much faster than did the strategies of the old public health. A number of key issues need to be brought closer to solution within ten years. Social innovation and political courage are called for (Bateson, 1975b): 'the ecological ideas implicit in our plans are more important than the plans themselves, and it would be foolish to sacrifice these ideas on the altar of pragmatism'.

References

The Adelaide Recommendations: Healthy Public Policy, 1988, *Health Promotion*, 3(2): 183–186.

Antonovski, A., 1979, *Health, Stress and Coping: New Perspectives on Mental and Physical Well-being*, Jossey-Bass, San Francisco.

Badura, B. and Kickbusch, I. (eds) (in press), *The New Social Epidemiology*, WHO Regional Office for Europe, Copenhagen.

Bateson, G. (ed.), 1975a, *Steps to an Ecology of Mind*, Ballantine Books, New York.

Bateson, G., 1975b, Ecology and flexibility in urban civilisation, in: Bateson, G. (ed.), *Steps to an ecology of mind*, Ballantine Books, New York, pp. 494–505.

Bateson, G., 1979, *Mind and Nature*, Fontana, London.

Beakhurst, G., 1979, Political Ecology, in: Leiss, W. (ed.), *Ecology versus Policies in Canada*, University of Toronto Press, Toronto.

Beck, U., 1986, *Risikogesellschaft, Augient Weg in Cine Underesiorierne*, Suhrkamp, Frankunamzlain.

Brewer, G.D., 1979, Policy Sciences, the Environment and Public Health, *Health Promotion*, 3(3): 227–237.

Bronfenbrenner, U., 1979, *The Ecology of Human Development: Experiments by Nature and Design*, Harvard University Press, Cambridge, MA.

Buttel, F.H., 1986, Sociology and the Environment: the Winding Road toward Human Ecology, *International Social Science Journal*, 109: 337–356.

Cassel, J.C., 1976, The Contribution of the Social Environment to Host Resistance, *American Journal of Epidemiology*, 104–107.

Catalano, R., 1979, *Health Behaviour and the Community: an Ecological Perspective*, Pergamon, New York.

Chamberlin, R.W., 1984, Strategies for disease prevention and health promotion in maternal and child health: the 'ecologic' versus the 'high risk' approach, *Journal of Public Health Policy*, 5: 184–197.

Dwivedi, O.P., 1986, Political Science and the Environment, *International Social Science Journal*, 38: 337–390.

European Monographs for Health Education Research, No. 6: 151–172.

Gleick, J., 1987, *Chaos: Making a New Science*, Penguin, London.

Green, L. and Raeburn, J., 1988, Health promotion. What is it? What will it become? *Health Promotion*, 3(2): 151–159.

Hancock, T., 1989, *Sustaining Health: Sustainable Development and Health*, Background paper for a workshop, York University, Ontario, June.

Hancock, T. and Perkins, R., 1985, The Mandala of Health: a Conceptual Model and Teaching Tool, *Health Educations*, 24(1): 8–10.

Illich, I., 1973, *Tools for conviviality*, Harper & Row, New York.

International Social Science Journal, 1986, Environmental Awareness (special issue) 38(3).

Kickbusch, I., 1986, Lifestyles and health, *Social Science and Medicine*, 22(2): 117–124.

Kuhn, T., 1970, *The Structure of Scientific Revolutions* (2nd edn), University of Chicago Press, Chicago.

Lalonde, M., 1974, *A New Perspective on the Health of Canadians*, Information Canada, Ottawa.

McKeown, T., 1976, *The Modern Rise of Population*, Edward Arnold, London.

McKeown, T., 1980, *The Role of Medicine: Dream, Mirage or Nemesis?* Princeton University Press, Princeton.

Mahler, H., 1988, Keynote address, *Health Promotion*, 3(2): 133–138.

Marmot, M.G., and Morris, J.N., 1984, The social environment, in: Holland, W.W. *et al.* (eds), *Oxford textbook of public health*, Vol. 1, Oxford University Press, Oxford.

Martin, C. and McQueen, D., 1989, *Readings for a New Public Health*, Edinburgh University Press, Edinburgh.

Mesarovic, M. and Pestel, E., 1974, *Mankind at the Turning Point*, second report to the Club of Rome, Dutton Readers Digest Press, New York.

Milio, N., 1983, *Promoting Health through Public Policy*, Davis, Philadelphia.

Milio, N., 1987, Making Healthy Public Policy; Developing the Science by Learning the Art: an Ecological Framework for Policy Studies, *Health Promotion*, 2(3): 263–284.

Milio, N., 1988, *Making Policy: a Mosaic of Australian Community Health Policy Development*, Melbourne, Victoria Department of Community Services and Health.

Ottawa Charter for Health Promotion, 1986, *Health Promotion*, 1(4): iii–v.

Press, E., 1987, Earth's Unique Environment, *News Report of the National Research Councils*, 38(8): 13.

Raeburn, J. and Rootman, J., 1988, Towards an expanded health field concept: conceptual and research issues in a new era of health promotion, *Health Promotion*, 3(4): 383–392.

Reich, M.R., 1988, Social policy for pollution-related diseases, *Social Science and Medicine*, 27: 10.

Rosen, G., 1958, *A History of Public Health*, M.D. Publications, New York.

Schnaiberg, A., 1980, *The environment: from surplus to scarcity*, Oxford University Press, New York.

Szreter, S., 1988, The Importance of Social Intervention in Britain's Mortality Decline 1850–1914: a re-interpretation of the role of public health, *Social History of Medicine*, 1(1): 9–37.

Terris, M., 1987, Epidemiology and the Public Health Movement, *Journal of Public Health Policy*, 8(3): 315–329.

Weekend Australian, 1989, April 1–2, p. 3.

Webster's Third New International Dictionary, 1976, Merriam Webster, Springfield, MA.

Winslow, C.E.A., 1923, *The Evolution and Significance of the Modern Public Health Campaign*, C.T. Elliott Books, Northford.

WHO, 1981, *Global Strategy for Health for All by the Year 2000*, World Health Organisation, 'Health for All' series, No. 3, Geneva.

WHO, 1988, *Basic documents of the World Health Organisation*, 37th edn, World Health Organisation, Geneva.

WHO, 1989, World Health Assembly resolution, WHA 42.26, 19 May 1989, WHO's contribution to the international efforts towards sustainable development.

World Commission on Environment and Development, 1987, *Our common future*, Oxford University Press, Oxford.

World Watch Institute, 1989, *State of the World Report*, World Watch Institute, Washington, DC.

3 Health and conservation

Shared values

Dan W. Walton and Peter Bridgewater

Abstract

Conservation and health are defined and their common roots in cultural values of society are discussed. Some of the principal components of both, such as scale, distribution, change, diversity, information, non-linearity, ethics, institutional competence and management are considered. The importance of the quality of life in a broad non-taxon specific sense is emphasised. Cultural inheritance and its relationship to technology, as well as the impact of technology on the environment, are discussed. Human population in relation to conservation and health practices is noted. The relationships between social pathologies of people, their cultural practices and the environment are issues of great import. The inadequacies and flaws of commonly used economic measures, especially as indicators of the quality of the environment and human health, are presented. The concept of carrying capacity is discussed, as is the significance of social equity. Concern about health and the environment is concern about the relationships which exist between people and the rest of the biosphere and, while there have been isolated successes, people have generally handled these relationships poorly. The need to integrate more fully the goals of conservation and health ethics for a sustainable society is acknowledged. The quality of any life, any living system cannot over time exceed the quality of the environment.

The challenge is to make the basic attitudinal changes, shed the undesirable baggage and make sure that our values and priorities coincide.

The extent of communication between environmental and human health professionals has, in some instances, been excellent. The European Charter on Environment and Health can be taken as an example of success. The role played by international health organisations in the activities leading up to and following UNCED is disappointing. Whether this is an example of barriers supported by the operators of professional theme parks, bureaucratic apathy or some bizarre division of the 'spoils' (= hegemony), is not clear. The joint involvement of health and environmental establishments in improving environmental quality at the national level, we suspect, varies widely between countries. In Australia, the National Health and Medical Research Council has solicited public debate on the issues and continues to seek input. How well the various professional and political subcultures respond is yet to be seen and evaluated. The quality of all life in Australia hangs in the balance. Can we develop a national environment and health charter in the mould of the European Charter or a charter with neighbours in our region?

Introduction

The relationship between people and the rest of the biosphere, and how we manage that relationship, is of major concern. The subject matter is far from simple and the literature cited here is but a small portion of that available. Certainly, this commentary is not the first to recognise what must be considered an indivisible entity, the well-being of the biosphere and the well-being of the human population. Presented here, however, are some thoughts on health and conservation which have their origins in recent Australian experiences, not the least of which was the initiative taken by the Australian government to conduct a national inquiry (1990–1992) into ecologically sustainable development (cf. ESDIIR, 1992).

The integration of information about the environment and about health, Health in its *broadest* context, is necessary to the development of policies for social structure and function which will be sustainable over time and space.

Conservation and health

The term conservation is widely and variously used. If we are to communicate with clarity, we must understand this word and the implications of our use of it. The World Conservation Union (IUCN), the United Nations Environmental Programme (UNEP), the World Resources Institute (WRI), in consultation with the Food and Agriculture Organisation of the United Nations (FAO) and the United Nations Educational, Scientific and Cultural Organisation (UNESCO), offer the following definition of conservation in the *Global Biodiversity Strategy*:

> The management of human use of the biosphere so that it may yield the greatest sustainable benefit to current generations while maintaining its potential to meet the needs and aspirations of future generations: Thus conservation is positive, embracing preservation, maintenance, sustainable utilisation, restoration, and enhancement of the natural environment.
>
> (Courrier, 1992)

Conservation is, therefore, rooted in a subset of values drawn from broader cultural values; values that determine how we deal with nature (Seligman, 1989). National biodiversity strategies, rather than 'reinventing the wheel', would do well to use this definition as their basis.

Professional disciplines invariably produce sub-cultures which are then integrated into organisational sub-cultures. These are not without their value, but they can also be inhibitory. With notable exceptions, the interchange of information or utilisation of shared principles among ecologists and health scientists has not been extensive.

Traditional ecologists have not devoted much attention to people and their domestic plants and animals. Human 'disturbance' was avoided. Conventional health scientists have largely excluded the wider environments in which people and their domesticated species live.

In many of the current discussions about environmental issues at a national and international level, people are not often treated or regarded as part of the biosphere and certainly not as part of biodiversity! And, unfortunately, the concept of health in its broadest context is not fully appreciated as central to that discussion.

One of the definitions of health provided by *The Macquarie Dictionary* (1981) is 'freedom from disease or ailment'. The idea of 'well-being, which represents an expansion of the idea of "freedom from disease"', is increasingly the present view of health (Mackay, 1993). The term health, like the term conservation, is a statement of a subset of values derived from the broader values of our society. Within this framework are many fields of scientific inquiry, not all of which focus on *Homo sapiens*. We must not confuse the issues by misuse of the term 'health'. Living systems, human or otherwise, can be regarded as ecologically healthy when their inherent potential is realised, their condition is relatively stable and their capacity for self-repair when perturbed is preserved and minimal support for management is needed (Karr, 1990).

Ecology has been defined as the 'science of the interrelations between living organisms and their environment' and is concerned especially with groups of organisms and functional processes (Odum, 1959), a view expanded as 'a discipline that integrates organisms, the physical environment, and humans' (Odum, 1992). The public perception of the health sciences, on the other hand, sees them as focused on the state of well-being of individuals and is largely reactive in approach. Ecology and epidemiology are related fields of scientific inquiry, albeit with flexible boundaries.

The point has been made (Soule, 1986, 1991; Beissinger, 1990) that nature conservation is largely crisis driven. The various scientific disciplines of conservation are also vulnerable to what people, including scientists and especially those sectors of our society which control the purse strings, view as a crisis. Resources of all kinds may be directed towards a particular threat, sometimes without any reasonable appraisal of the importance or complexity of the threat relative to other needs (Harwell *et al.*, 1992).

Health sciences undoubtedly are equally vulnerable to such influences. Science disciplines, whether conservation or health oriented, are as susceptible to 'trendy' techniques, ideas and the pursuit of fame and fortune as any other part of society (Walton, 1994). The training and the work of science professionals, including associated management professionals, are conducive to linear, personal, goal-directed behaviour.

Living systems, on the other hand, regardless of their level of organisation, are simply not linear systems. The challenges that invite pursuit from such integrative considerations are to contain crises; to control the opportunity for conflict; to avoid seemingly simple linear approaches and solutions, by calling on a more integrative creativity and developing a social equity that seeks to allow for due recognition yet protect privacy.

There is the inevitable temptation to drift into a presentation demonstrating that basic ecological principles pertain regardless of species, community, ecosystem, landscape or seascape. Differences do lie in detail and subsets of ecological principles pertain to those details. No species except *Homo sapiens* is capable of a rational decision to live in an ecologically sustainable manner.

Any species when faced with an environmental crisis which threatens the success of its own perpetuation has four basic options. It may:

- relocate to a more friendly environment;
- attempt to remove or nullify the threat in its environment;
- not respond and 'hope' for the best; or
- become extinct.

The decision to try to live in an ecologically sustainable manner is perhaps an indication that we accept that there is no suitable place left to which we as a species may move and that we prefer to avoid extinction. Australians have seemingly selected the second option, but the third could emerge by default. How we define sustainability – a reflection of the values of our society – will inform what functions we will attempt to determine if we are reaching our goal of sustainability. The analysis of data gathered on those indicator functions will determine how we measure the health of our landscapes, seascapes, ecosystems, communities, species and individuals.

In the development of perspectives (Ferguson, 1994) for healthy landscapes, seascapes, ecosystems, communities, species and individuals, there will always be the need to cater for the extremes: the intensive care for the seriously ill, whether people or endangered species, communities, ecosystems, landscapes and seascapes. Soil quality is the very basis of sustainable land use (Thomas and Kevan, 1993). A healthy flora is critical to a healthy fauna. Water and air quality influence the quality of all life. Mental health is inextricably linked to physical health and mental health is a complex integrative aspect of life which draws heavily on and determines the values of our society (Parry-Jones, 1990; for an interesting analogy, see the comments on the Last Straw Syndrome by Mackay, 1993).

Prevailing values derived from the current beliefs of society can be influenced and shaped over time by information scientifically gathered, but at any given moment those values and beliefs are more important in the shaping of public policy than the results of the latest scientific research. Science is the application of a method of inquiry. Inquiries produce information. Information derived from scientific inquiry can be used to solve problems, but problems are identified by people, are defined by people and people attempt to solve problems. Identification, definition and solution are constructs derived from value systems of human society, just as indeed is the awareness of the opportunities for inquiry and the methods to be employed.

Not only has no species besides *Homo sapiens* been capable of making the decision to live in an ecologically sustainable manner – never before has any species been sustainable ecologically. If ecological sustainability is truly the agreed goal of Australians (ESDIIR, 1992), the role of science in reshaping and influencing the values and beliefs of Australian society is absolutely critical to the achievement of that goal. Ecological sustainability is ecologically sustainable prosperity for the entire biosphere. Prosperity, as the word is used here, should not be confused with financial wealth!

Principal components

Whether the issue is health or conservation, there are principal components which must be addressed as central to the development of all strategies and principles. There are various terminologies which may be employed, most often a reflection of perspective rather than content. Some of the components are presented below, especially those considered highly relevant to the establishment of priorities.

Scale

One of the significant problems in any discussion of health or conservation is to maintain an awareness of scale and of the existence of more than one scale (Daly, 1991). In purely people terms, this is the confrontation of the concern of and for the individual versus those of and for the group, balancing self-interest with wider interests.

For conservation, the concerns can be expressed as populations versus landscapes and seascapes. In the final analysis, there are legitimate and proper concerns at all scales. Conflict arises in the establishment of priorities and the allocation of resources. Most issues can be selectively and variously scaled, depending upon the associated values and beliefs.

Distribution

Biotic systems and abiotic features are rarely, if ever, distributed evenly or randomly in time or space. Nor are they determined by political boundaries. Highly dynamic patterns and specific associations do exist. Most of the global human population, certainly the Australian population, lives in the coastal zone close to the mix of plentiful freshwater, marine resources and arable lands. Values and beliefs, also, are dominant ingredients in human distribution patterns.

Change

All biotic systems and abiotic features change in time and space. Human society, however, deals poorly with change and devotes many resources to the denial of change and ill-fated attempts to resist change (Walton, 1994). The costs of this mal-adaptation are huge and not confined to people. Many of the institutions of society cultivate corporate cultures which gradually divert more and more resources into resisting change. Those who work closely with change, including health and conservation scientists, are potential captives of, or contributors to, this mal-adaptation.

Human influence is widely distributed throughout the biosphere. Much has been shaped, directly and indirectly, by human activity. One must be very cautious, therefore, in the use of the word 'natural' (Sprugel, 1991). Through intervention, active or passive, we may direct and influence the rate of travel, but we cannot arrest time's arrow.

Diversity

There are three basic and interactive elements of diversity: cultural, biological and place. The importance of all three elements should not be minimised, nor should one be allowed to dominate. Human identity is derived from the intellectual interpretations of the interactions of these elements.

Information

The absolutely fundamental construct must be that knowledge is better than ignorance. Information access should be unrestricted and education an ongoing process in a free and democratic society (Miller, 1994). Quality standards and ease of access to information are essential if there is to be social equity and if we are to manage the biosphere for ecological sustainability (Busby and Walton, 1994).

Non-linearity of problems

There is always the attempt to simplify a problem for this is the prerequisite to a simple solution. Such simplification fits neatly with many of the institutional arrangements of human society, but biotic systems are highly complex and, for that matter, because it is such a system, so is human society. The acquisition of information may be achieved via a reductionist approach, but problems must be addressed in a more holistic manner if sustainability is desired. Small, isolated decisions represented as simple solutions invariably accumulate as very large problems whose complexity is increased by the store of undesirable outcomes from those small decisions (Odum, 1982).

Ethical framework

In the value systems of all societies there is a framework of ethics. Ethical issues, non-economic and non-scientific, are invariably charged with emotion and tend to have an importance which cannot accurately be given a monetary worth. They are often elusive to adequate definition, yet are the adhesive for the fabric of society. To ignore this framework, to fail to understand its importance, is to preclude the possibility of ecological sustainability. The expression of the essence of the ethical framework is embodied in the word quality. If ecological sustainability is prosperity, society will ask about the quality of that prosperity. Quality must begin with the common stewardship of the land, the waters and the air.

Institutional competence

Human institutions are invariably created with the best of intentions, to accomplish a definite, if not noble, purpose and in a burst of enthusiasm that captures the values of its creators. All institutions, however, age; the founding intentions are forgotten or appear no longer relevant. The purpose seems to fade and there are value shifts. Institutions foster subcultures which, in some cases, are transformed into icons (customs and traditions). At some point, many become bizarre theme parks where the citizens, customers, patients, inmates, resources, or whatever, become the means by which the members of the subculture flourish with no thought beyond the perpetuation of that sub-culture (Enckell, 1982). Institutional renewal and replacement are social imperatives.

Management

Among scientific and health related organisations, management has not been generally perceived as an important and contributing element. Management acquired a reputation as the province of 'those who couldn't make it' as scientists, or accountants with unwarranted ambitions. Consequently, results have been less than desirable.

The view that *Homo sapiens* is in constant battle with the remainder of the biosphere has inculcated the contrasting images of people as the conqueror and people as the victim. Intentionally or otherwise, people have altered the biosphere and to be a sustainable part of it, management rather than confrontation is required (Bridgewater *et al.*, 1992; Walton, 1994).

Ecology and the environment

As noted previously, ecology is a field of scientific inquiry and, as such, is committed to the search for information which may be synthesised into principles. Environmentalism, however, refers to the movement, the aggregate, of those people concerned about the quality of the environment. This is an important distinction about which to be clear: the difference between the search for information on the one hand and the application of information in the pursuit of environmental quality on the other. Ecology sits uneasily with what Western economists view as development (Geerling *et al.*, 1986) and environmentalism has not fully come to grips with people as a species within the biosphere (Grizzle, 1994).

Whether the word apathy, or perhaps a more pejorative term, is appropriate in this particular context, it is interesting that an adequate environment (adequate for human health and well-being) is not enshrined as an inherent right of every citizen. Only nine countries (Brazil, Chile, Ecuador, Nicaragua, Panama, Peru, Portugal, the Republic of Korea and Spain) include this right explicitly in their national constitutions (Westing, 1993). While this list may well provoke some cynicism, the brevity of the list is cause for serious concern.

Ecological research, like all scientific research, is about the acquisition of information. Like every other cultural expression of human society, scientific research has accumulated a mythology. This is not the place to review all aspects of this mythology, but some points must be made. There are those within the community of research scientists who set great store by fostering the idea that scientific research may be divided into 'basic' science and 'applied' science. This false dichotomy is more closely related to attempted hegemony of the research agenda than any other reality. A much more appropriate division would be 'curiosity driven' research and 'mission oriented' research. All science is carried out by people for people. Scientific research should not be perceived by the public as subsidised self-indulgence, nor must intellectual freedom be equated with social irresponsibility (Fox, 1994; Santana and Jardel, 1994). Public expectations, including those of government, of scientific research must be realistic (Aitkin, 1992; Richmond, 1993).

In an attempt to integrate, at least in part, ecology and environmentalism through the reform of the federal environmental research and development programmes in the United States, there is proposed the establishment of a new agency, the National Institute for the Environment (Howe, 1993). The new agency would function along three research lines: what do we have? (inventories, monitoring and characterisation), how does it work? (mechanisms, processes and effects) and how do we maintain it? (strategies, technologies and solutions).

Existing federally funded environmental research in the United States was found to have five major flaws, presented here in brief:

- funding was not commensurate with need;
- leadership was weak or non-existent;
- the research itself was not as effective as it should be;
- environmental research was heavily skewed towards physical sciences; and
- feedback between assessment of societal needs and research priorities was insufficient.

Does a similar situation pertain more widely? What is the present status of environmental research? There is reason to believe that all five flaws also characterise research, including environmental research, in Australia (Aitkin, 1994). Recommendations were made regarding environmental research in Australia (ASTEC, 1990); the impact of those recommendations should be assessed and their continued appropriateness evaluated.

People

The biological species

Homo sapiens possesses characteristics which are unique, but this can be said for every other biological species as well as a variety of molecular aggregations on the margin of what we define as a living organism. The human species is an animal, a chordate, a vertebrate, a mammal, a primate and a hominid. All of these terms characterise our species biologically. There are, of course, many other forms and characteristics which define our species. And, like any other species, we have an ecology – we interact with all other elements of our outside environment and respond to interactions of our individual internal environments.

The interactions of and with our environment are highly complex. Cells, tissues, organs and individuals interact. There are species of plants and animals which parasitise us, some within our bodies and others on the body surface. There are many species, plants and animals, which are commensals living in our clothing, furnishings, foodstuffs, books, habitations or other associated places or things. Some species of plants and animals have been attuned to specific human needs, such as domesticated species, and they bring their associated parasites and commensals also to live with people.

The diversity of species with which we interact on a more or less intimate basis is very great and the frequency of contact extends from those which are virtual permanent residents to those with which any contact may be the last.

We rely heavily upon the 'free' services of other species to carry out tasks. These include the formation of foodstuffs; the disposal of wastes and the cycling of nutrients and minerals; the decontamination of soil, water and air; assistance in the regulation of ambient temperature and humidity; the control of ground water levels and the prevention of erosion as well as the provision of a wide array of aesthetic qualities.

Consideration of humans as a biological species (Diamond, 1991) and subject to the broad ecological principles applicable to other species is not without opposition (Grizzle, 1994), especially from those who would view humans as somehow of extra-terrain origin. Nevertheless we prey on other species (carnivory and herbivory) for food and other resources. We compete with other species for resources. In a broader sense we may be a formidable parasite (Odum, 1992), a commensal, a symbiont or the most successful coloniser in evolutionary history, the archetypical superweed (Cohen and Stewart, 1994). And, like other species, we alter the place where we live and are quite capable of making that place uninhabitable. The more of the total available energy is used by people, the less is available to other species.

The species with culture

The human species differs from most other species, except perhaps primates and cetaceans, because people have culture: what the *Macquarie Dictionary* defines as 'the sum total of ways of living built up by a group of human beings, which is transmitted from one generation to another'. (For some interesting observations on Australian culture vis-à-vis the environment, see Dunlap, 1993). While the ability to pass non-genetically coded information from one generation to another can be viewed as an asset to the species, this ability also increases the ecological complexity of the species.

Two serious and relevant complications are the ability to use and store information for the purpose of solving problems (developing technology) and the diversity of cultural values which is possible. Neither technologies (even if health oriented) nor cultural values are environmentally benign, and any meaningful study of human ecology must deal with the consequences of technologies and prevailing values (Cramer and Zegveld, 1991; Ludwig et al., 1993; Nierenberg, 1993; Stern, 1993).

The existence of technologies and cultural values does not imply sustainability. Every action consumes resources and produces wastes. Indeed, changes in cultural values are historically associated with military conquest, religious conversion, shifting political elites and technological shifts. A period of deterioration and destabilisation of prevailing environmental conditions appears invariably to accompany cultural change. The goal of sustainability implies a major shift away from being a crisis-driven, brink-of-disaster society to long-term viability centred on life-oriented goals (Cohen and Polunin, 1990).

The cultural values of any society attempt to capture quality, that attribute which is so elusive to description yet so incredibly distinctive. In our attempt to become a viable sustainable society, we must not fall into the trap of replacing reality with symbols. Zoos, botanical gardens, hospitals, intensive care units, nursing homes, day-care centres, research centres, universities, court rooms, rehabilitation centres, and other places are value laden and are some of the symbols of what we regard as quality of life. Such human institutions cannot replace the wider conservation reality required to attain the quality of life for a viable sustainable society. Existing environmental quality must be maintained, damage prevented and faults repaired (Kuusinen et al., 1994).

Human population

The matter of human population may be considered as a biological characteristic or as a cultural phenomenon. The interaction of biology and culture causes a divergence from those ecological principles which apply to other species, regarding circumstances relating not only to natality, morbidity and mortality, but also to those influences upon the internal environment of individuals and external environmental influences on individuals and various tiers of human aggregations, circumstances and influences which feed back into biological and cultural attributes.

Global population level is a major issue considered by various international fora. Major concern has been expressed, not only about the total, but about the rate of growth, of the human population (TRSNAS, 1992; UN, 1992; Myers, 1993;

Bongaarts, 1994). Speculation on the relationship of the human population size to the carrying capacity of the biosphere has resulted in reference to 'demographic winter' (Daily and Ehrlich, 1992). Poverty is a cause and effect of environmental problems and there is a complex relationship between poverty, population and environmental degradation (Lonergan, 1993). To imply that environmental degradation and the socially disadvantaged are independent (unrelated) phenomena is disquieting. There can be no doubt that human population growth and growing human aspirations are on a collision course (Walton, 1994). Wilson (1988) terms this the essential paradox of human existence: the drive towards perpetual expansion or personal freedom. On a local scale, the control of resources as a means of retaining or attaining a perceived cultural superiority can be a factor in the manipulation of the population level (Shaw, 1993).

Among relevant cultural values involved in any consideration of human population are those values related to religion and ethical or moral precepts. Such issues are socially highly volatile. Reason quickly gives way to emotion and otherwise sensible debate is finished. Dealing with these values is further complicated by the contrast in what people profess and what they actually do. Institutional structures and corporate cultures become involved. Full and open discussion of human population issues is seldom possible, but the ramifications of the rapidly increasing population for the ability to be a sustainable species are monstrous (Smil, 1993). The implications for the epidemiology of communicable diseases, not only of people but of their associated domesticated, commensal, peri-domestic and parasitic species, are staggering (Krause, 1992; Bos, 1994). Movements of large numbers of people in response to environmental change, including disease and contamination, already pose serious problems (Kennedy, 1993; Westing, 1994).

True, Australia has a population of some 17 million people, but there are somewhere around 300 million domestic animals (including ferals). Food, fibre and fodder crops support these populations. Westing (1993) points out that the available land on Earth per caput in 1970 was 3.6 hectares and that by 1990 there had been a 30 per cent decline to 2.5 hectares. If we also attempt to estimate the available land for domestic animals, we come to some startling conclusions about where we may be standing with regard to prevailing environmental health conditions. We also begin to gain some appreciation of the stress on resources, especially if we are reminded that people (and their domesticated animals) are not distributed evenly or randomly across the land surface. While there are connections between human population growth rate in, and the affluence of, specific societies, there can be no truthful denial of the fact that at the global level available land surface per caput is decreasing.

Discussion

The view that health problems are what are made manifest in the rooms of a clinician belongs in the junk pile of history. Unfortunately, such views persist disguised as rather emotive pleas for traditional research funding (for example, see Kirschner *et al.*, 1994). The full range of economic and other social stresses is manifest among individuals, families, communities and nations as structural and functional pathologies. Inappropriately planned development and reckless alterations to the environment are invitations to unpleasant surprises across a wide spectrum of health concerns. There is no tradition of monitoring the effects of development or environmental change. The narrow focus on illness challenges in the clinical arena effectively masks the broad hazards of environmental toxicity, e.g. neurotoxicity, reproductive toxicity, respiratory toxicity.

Public consultation usually involves little more than the confrontation of the few vested interests who understand the issue(s) and hope thereby to gain or defend their interest(s). Genuine public consultation is most often regarded as too hard, for it requires that people have the issues explained to them in some detail and their informed views solicited. Open, democratic processes are not really cultivated at the range of scales for effective functioning as a free and open democratic society. Whether traditional health professionals have contributed to open debate, at least on some issues, has been challenged (Gaze, 1992).

Consideration is being given to improving the way industry operates (Geiser, 1991) and to the development of regulatory mechanisms (Gunningham, 1994). The traditional policy approach to control over processes is prohibitory: the detailing of what is prohibited with a corresponding menu of punishments for violators. Prohibition is a notorious failure as an implemented policy approach. Enforcement is highly vulnerable to political distortions of all kinds. The preferred option is to make compliance rewarding and the failure to comply a punishment in itself.

The central focus should be on risk assessment, risk prevention, risk remediation and long-term liability. Policy options in these areas will largely be adventures in regulatory strategy. Impacts of development and environment on health are multi-dimensional and exist in time and space. We need to honour the precautionary principle and have the appropriate regulatory strategies in place, the necessary monitoring systems working, and appropriate preventive and curative 'health' care delivery systems operative.

In parallel to all this is the need to alter radically the prevailing economic structure; to understand and take account of the full costs of resources used (Hubbard, 1991), costs of replacement of resources for future uses and full costs of produced wastes. Soundness and freedom from disease or damage must be recorded as a resource (Hall et al., 1992).

The conventional and widely used measure of national wealth (GDP) has serious flaws. Three of the most serious are that: only economic activity which involves money-based transactions is considered; use of such a restricted measurement makes it difficult to allow for the quality of life (e.g. costs of crowding, noise, air pollution); and behaviour (e.g. court cases) which only redistributes resources is counted (McRae, 1994).

New policies, technologies and developments are a reality and they will continue, but too little attention has been given to their introduction in such a way that each increment is reversible. Similarly, as we noted above, there is no tradition of monitoring the impacts of newly introduced policies, technologies and developments

(Walton *et al.*, 1992). Assessments of 'potential' impacts are seldom measured against subsequent real impacts, and the hypothetical is seldom compared with reality (Machlis, 1992). Ecology offers an alternative perspective to money in the assessment of economic and social activities, including impacts upon future generations (Hawken, 1993).

No doubt there needs to be a major review of responsibility and accountability in regard to a wide range of environmental matters. This, however, opens very significant issues regarding land tenure (Houghton, 1994), roles of the various spheres of governments, the appropriateness of various institutions of society, a variety of legal processes and a host of other issues including the responsibilities and liabilities of public officials for outbreaks of preventable diseases or preventable environmental hazards. The relationships between economic and social responsibility are part of the equity essential to sustainability (Sarokin and Schulkin, 1992).

Social change and shifts in political paradigms must be considered (Giampietro and Bukkens, 1992; Macrae *et al.*, 1993). The gap between what is professed and what is produced must be narrowed. While most societies pay lip service to concern about future generations, the crisis-of-the-day approach is more the reality. Governments are elected and serve for fixed or variable durations. During this period, the basis for re-election must be established (or cause for defeat prevented). Short term planning and thinking prevails, in part dictated by the crisis re-election syndrome.

Legislative protocols for long term concerns have been most effective where emphasis has been on prevention (e.g. prevention of the so-called childhood diseases, safe water supply, at least partially adequate sewage and waste disposal).

Perhaps encouraged by scientists seeking funding, governments have tended to look for and expect to find magic bullets – one-shot miracle cures for problems. On-going social, economic and environmental problems either fall into the 'too hard' basket or are not addressed in the context of health. To what extent professional boundary disputes are a barrier to satisfactory handling of such issues is not known.

Society has been very slow to credit the clear and definite connection of population growth, increased demands for food, fuel and living space and environmental degradation to what is termed here as 'health, well-being, safety and comfort' (Gowdy, 1992). Scientists have not been highly successful, perhaps not very energetic, in directing public attention onto this holistic view (Abelson, 1994), but rather have left the struggle to those ethically committed but scientifically naive. The transfer of research derived information to policy formulation and implementation occurs at rates which are inversely proportional to the perceived threat of change from 'business as usual', a threat exaggerated by the lack of a clear understanding of what it means to be environmentally conscious (Kleiner, 1991).

There must be a realistic understanding of what resources are available for development assistance, what can reasonably be expected of such resources and what priority is given to where and on what these resources can be utilised. Similarly, a careful assessment is needed of how well international agencies, such as WHO, are performing and in which areas bilateral assistance may best contribute. What contribution can be made to issues under consideration by the UN Commission on Sustainable Development and what priority should these issues be accorded?

The opportunity exists to address large scale problems through the formation of partnerships. Such associations can amplify the effect of resource input and deal with problems of mutual interest and concern (Goldemberg, 1993). An interesting initiative is in progress at the moment, an attempt to form a Southern Hemisphere consortium of nations which can bring the southern perspective to global issues. Such efforts and alliances should be fostered.

One must be cautious, however, in development assistance. Worker safety, conventional public health and environmental standards are easily sacrificed in the name of cost-savings. Reducing costs by lowering standards and reducing costs by increasing efficiency are very different pursuits with vastly different end products (Daly, 1992). The view that economic based growth is the best way to spread economic opportunity and social well-being is a highly suspect linear rationalisation (Abelson, 1993).

Attempts to discuss human population and settlement policies in light of local, regional and national carrying capacities not only address very fundamental aspects of human ecology (Ophuls and Boyan, 1992), but also reveal the depth of our ignorance of human ecology. We must admit that most human demography is a product of almost everything except the utilisation in planning and management of information derived from scientific research. Such research would take into account that people are social and technological mammals and are further characterised by a cultural as well as a genetically based inheritance. Whatever standards exist which are related to population and settlement have been largely historically derived through religious beliefs and cultural conflicts. In the modern democratic political process, mythology and fact are variously valued, the former frequently outweighing the latter (Kennedy, 1993; McRae, 1994).

Modern technology and scientific advances have raised human expectations and aspirations to great heights. Many of these developments have revealed an unanticipated price. Initial successes may bring on ultimate crashes as represented by such things as air pollution from automobiles, radioactive waste, resistance to pesticides, resistance to antibiotics, salinisation and waterlogging from irrigation practices. Human activities have exerted an influence on sea level (Sahagian *et al.*, 1994) and global effects of land use change are emerging (Houghton, 1994). That the expanding human population, and increasing urbanisation, is an expanding medium of opportunity for microbial evolution should not be doubted (Krause, 1992; Hughes *et al.*, 1994).

A clear understanding of the carrying capacity (Carey, 1993) of the major landscapes of Australia, the optimum size and distribution of communities and a knowledge of sustainable growth rates is badly needed. At present, the discussions regarding the desirable population size for Australia are driven largely by emotions, not all of which are admirable. There is a dearth of good, detailed analyses of carrying capacities and environmental costs. We must, however, avoid the trap of increasing spending on environmental matters (or ecology) as a front for the failure to confront population issues (Ludwig *et al.*, 1993).

We must agree on what is meant by health. Having made that point, no serious consideration of a sustainable society can take place without provisions for efficiency (minimisation of waste) and social equity. Social equity is not an ideological form of government, such as communism, as some think, but a value system of a society which accords real value to all people and to all places.

Either all of Australia and all Australians are important or we risk assigning all to irrelevance. This value system does not preclude surgeons and janitors, but it should preclude toxic waste dumps and the use of urban streets as mental health care centres. Allocation and access are only one side of the issue; there is also the responsibility to ensure that allocation and access are within the sustainable limits of the resources (Upretii, 1994).

Although the word has been trivialised by usage, people must be 'empowered' by having easy access to information and acquire the sense of not just belonging but of being essential to the operation of society (Vittachi, 1989). This question should not be a debate about powers of competing tiers of government, and community should not be restricted by archaic lines on maps which were established without regard to ecological or cultural information. Planning will be imperative and the financial sector must learn to deal with non-economic values now given little credence. Education of the community is critical; trades and professions will have to share information now viewed as sources of power for the trade or profession. Efficiency will demand a reassessment of many public and private institutions and many archaic and antiquated customs and traditions must also be evaluated for effectiveness.

There are existing programmes coordinated by international organisations, such as the Man and the Biosphere Programme (MAB) handled by UNESCO, which lend themselves very well to sustainable community development and the enhancement of democratic processes.

Conclusion

From this discussion, two central points emerge:

- concern about health and environment is really concern about the relationships which exist between people and the rest of the biosphere; and
- while there have been isolated successes, on the whole people have not managed effectively their interaction with the environment and a situation exists with the potential to get out of control.

So, what is to be done?

If we are to be a sustainable society, the minimum condition required is that we allow no decline of our natural capital stock below its current level (Costanza and Daly, 1992). We must, in fact, follow the prescription implicit in the definition of conservation given herein and enhance that capital stock.

The quality of that stock is, in part, determined by its health. No doubt we will need measures for quality and standards as guides to quality. Research, in itself, will not prevent the decline of natural capital, will not enhance existing stock or rehabilitate that which is diseased or injured. Information is not an end product. Research results must be integrated into management, the results of management strategies monitored and new research priorities established on the basis of management needs. Barriers presented by professional and corporate subcultures must be broken down. Much more attention should be given to the education of potential managers (see McNeely, 1995).

Human population growth rates are historically related to technological breakthroughs or the release of previously unavailable resources. Science and technology must not raise false expectations that discoveries or developments will provide miracle cures or solutions to huge and intractable problems (TRSNAS, 1992).

Accountability and liability must take on new dimensions. The assignment of a monetary value to everything ascribes a value to money beyond the aesthetic values of quality and the ethical constructs of society. While every society appears to need a mythology, this mythology should be benign at worst and constructive at best. Infirmity and damage are not restricted to physical systems, but extend to the behavioural as well. It is in this area – changes in human behaviour – that real advances in human and environmental health will occur (McRae, 1994).

What goals have been declared for our society, what are we doing to achieve these goals, what is the rate of progress towards the goals and are we constantly updating the goals, methods and rates?

Environmental reconstruction is quite likely to be a dominant economic and social force in the future, especially if society makes the commitment to ecological sustainability, just as new combinations and new associations are likely to dominate ecology. The broader goals of conservation will provide the framework. The famous bottom line of business may be the point where imagination ceases and mental stagnation takes over (Walton, 1994). There is little positive transfer from the awe of spectacular scenery to building what Caldwell (1994) terms sanative communities. Nevertheless, the quality of the landscape and the quality of life are intimately related. Any society accumulates antiquarian junk, parts of which are inappropriate and, perhaps, socially and ecologically harmful.

The challenge is to make the basic attitudinal changes, shed the undesirable baggage and make sure that our values and priorities coincide.

The extent of communication between environmental and human health professionals has, in some instances, been excellent. The European Charter on Environment and Health can be taken as an example of success. The role played by international health organisations in the activities leading up to and following UNCED is disappointing. Whether this is an example of barriers supported by the operators of professional theme parks, bureaucratic apathy or some bizarre division of the 'spoils' (= hegemony), is not clear. The joint involvement of health and environmental establishments in improving environmental quality at the national level, we suspect, varies widely between countries. In Australia, the National

Health and Medical Research Council has solicited public debate on these issues (NH&MRC, 1991). How well the various professional and political subcultures respond is yet to be seen and evaluated.

The quality of all life in Australia hangs in the balance.

Can we develop a national environment and health charter similar to the European charter or a charter with neighbours in our region? Can a positive strategic plan emerge, one which emphasises those things which yield results that improve or maintain quality living systems? There is much more to life than morbidity and mortality.

References

Abelson, P.H., 1993, Policies for science and technology, *Science*, 260: 735.

Abelson, P.H., 1994, Science, technology, and congress, *Science*, 263: 1203.

Aitkin, D., 1992, Aligning research with national purpose, *Canberra Bulletin of Public Administration*, No. 68: 110–115.

Aitkin, D., 1994, In search of a national research plan, *R&D Review*, February: 25–26.

ASTEC, 1990, *Environmental Research in Australia: The Issues*, A Report to the Prime Minister by the Australian Science and Technology Council, Australian Government Publishing Service, Canberra.

Beissinger, S.R., 1990, On the limits and directions of conservation biology, *BioScience*, 40: 456–457.

Bongaarts, J., 1994, Population policy options in the developing world, *Science*, 263: 771–776.

Bos, L., 1994, Environment and disease: evergrowing concern, *Environmental Conservation*, 21: 99–102.

Bridgewater, P.B., Walton, D.W., Busby, J.R. and Reville, B.J., 1992, Theory and practice in framing a national system for conservation in Australia, in: *Biodiversity: Broadening the Debate*, Canberra, Australian National Parks and Wildlife Service: 3–16.

Busby, J.R. and Walton, D.W., 1994, A national biological survey for the United States? Comparable Australian activities at the national level, in: Longmore, R. (ed.), *Biodiversity – Broadening the Debate*, 3, Australian Nature Conservation Agency, Canberra: 4–11.

Caldwell, L.K., 1994, Landscape, law and public policy: conditions for an ecological perspective, *Landscape Ecology*, 5: 3–8.

Carey, D.I., 1993, Development based on carrying capacity: a strategy for environmental protection, *Global-Environmental Change*, 3: 140–148.

Cohen, J. and Stewart, I., 1994, *The Collapse of Chaos: Discovering Simplicity in a Complex World*, Viking, London.

Cohen, P. and Polunin, N., 1990, The viable culture, *Environmental Conservation*, 17: 3–6.

Costanza, R. and Daly, H.E., 1992, Natural capital and sustainable development, *Conservation Biology*, 6: 37–46.

Courrier, K. (ed.), 1992, *Global Biodiversity Strategy: Guidelines for Action to Save, Study, and Use Earth's Biotic Wealth Sustainably and Equitably*, WRI, IUCN, UNEP, Washington DC.

Cramer, J. and Zegfeld, W.C.L., 1991, The future role of technology in environmental management, *Futures*, 23: 451–468.

Daily, G. and Ehrlich, P.R., 1992, Population, sustainability, and Earth's carrying capacity, *BioScience*, 42: 761–771.

Daly, H.E., 1991, Towards an environmental macroeconomics, *Land Economics*, 67: 255–259.

Daly, H., 1992, Free trade, sustainable development and growth alert: some serious contradictions, *Ecodecision*, No. 5 (June): 10–13.

Diamond, J., 1991, *The Rise and Fall of the Third Chimpanzee*, Vintage, London.

Dunlap, T.R., 1993, Australian nature, European culture: Anglo settlers in Australia, *Environmental History Review*, 17: 25–48.

Enckell, P.H., 1982, Development and strategies of scientific institutes, *British Ecological Society Bulletin*, 13: 16–20.

ESDIIR, 1992, *Ecologically Sustainable Development Working Group*, Intersectoral Issues Report, Australian Government Publishing Service, Canberra.

Ferguson, B.K., 1994, The concept of landscape health, *Journal of Environmental Management*, 40: 129–137.

Fox, C.H., 1994, If it ain't fixed, don't break it . . ., *Nature*, 369: 602.

Giampietro, M. and Bukkens, S.G.F., 1992, Sustainable development: scientific and ethical assessments, *Journal of Agricultural and Environmental Ethics*, 5: 27–57.

Gaze, B., 1992, Controlling medical science: reproductive technology, infertility and the position of women, in: Arup, C. (ed.), *Science, Law and Society*, La Trobe University Press, Melbourne, pp. 29–55.

Geerling, C., Breman, B. and Berczy, E.T., 1986, Ecology and development: an attempt to synthesise, *Environmental Management*, 13: 211–214.

Geiser, K., 1991, The greening of industry: making the transition to a sustainable economy, *Technology Review*, 94(6): 64–72.

Goldemberg, J., 1993, The geopolitics of environmental degradation, *Environmental Conservation*, 20: 193–194.

Gowdy, J.M., 1992, Economic growth versus the environment, *Environmental Conservation*, 19: 102–104.

Grizzle, R.E., 1994, Environmentalism should include human ecological needs, *BioScience*, 44: 263–268.

Gunningham, N., 1994, Developing an optimal regulatory strategy, *Search*, 25(4): 98–101.

Hall, J.V., Winer, A.M., Kleinman, M.T., Lurmann, F.W., Brajer, V. and Colome, S.D., 1992, Valuing the health benefits of clean air, *Science*, 255: 812–817.

Harwell, M.A., Cooper, W. and Flaak, R., 1992, Prioritising ecological and human welfare risks from environmental stresses, *Environmental Management*, 16: 451–484.

Hawken, P., 1993, *The Ecology of Commerce: How Business can Save the Planet*, Weidenfeld & Nicolson, London.

Houghton, R.A., 1994, The worldwide extent of land-use change, *BioScience*, 44: 305–313.

Howe, H.F., 1993, The national institute for the environment: comparison with other proposals, *The Environmental Professional*, 15: 428–435.

Hubbard, H.M., 1991, The real cost of energy, *Scientific American*, 264(4): 18–23.

Hughes, J.M., Peters, C.J., Cohen, M.L. and Mahy, B.W.J., 1994, Hantavirus pulmonary syndrome: an emerging infectious disease, *Science*, 262: 850–851.

Karr, J.R., 1990, Biological integrity and the goal of environmental legislation: lessons for conservation biology, *Conservation Biology*, 4: 244–250.

Kennedy, P., 1993, *Preparing for the Twenty-first Century*, Fontana, London.

Kirschner, M.W., Marincola, E. and Teisberg, E.O., 1994, The role of biomedical research in health care, *Science*, 266: 49–51.

Kleiner, A., 1991, What does it mean to be green? *Harvard Business Review*, 64: 38–47.

Krause, R.M., 1992, The origin of plagues: old and new, *Science*, 257: 1073–1078.

Kuusinen, T., Lesperanoe, A. and Bilyard, G., 1994, Toward integrated strategies for achieving environmental quality, *The Environmental Professional*, 16: 22–27.

Lonergan, S.C., 1993, Impoverishment, population, and environmental degradation: the case for equity, *Environmental Conservation*, 20: 328–334.

Ludwig, D., Hilborn, R. and Walters, C., 1993, Uncertainty, resource exploitation, and conservation: lessons from history, *Science*, 260: 17–36.

Machlis, G.E., 1992, The contribution of sociology to biodiversity research and management, *Biological Conservation*, 62: 161–170.

Mackay, H., 1993, *Reinventing Australia: The Mind and Mood of Australia in the 90s*, Angus & Robertson, Sydney.

McNeely, J.A., 1995, *Strange Bedfellows: Why Science and Policy Don't Mesh, and What Can Be Done About It*, Presentation to 'Living Planet in Crisis: Biodiversity Science and Policy', American Museum of Natural History, New York City, 9–10 March 1995, Typescript, 11 pp.

Macrae, R.J., Henning, J. and Hill, S.B., 1993, Strategies to overcome barriers to the development of sustainable agriculture in Canada: the role of agribusiness, *Journal of Agricultural and Environmental Ethics*, 6: 21–51.

McRae, H., 1994, *The World in 2020, Power, Culture and Prosperity: A Vision of the Future*, HarperCollins, London.

Miller, R.B., 1994, The role of information in public policy, in: Longmore, R. (ed.), *Biodiversity – Broadening the Debate*, 3, Australian Nature Conservation Agency, Canberra, pp. 12–20.

Myers, N., 1993, The 6 'biggies', *People and The Planet*, 2(4): 31.

National Health and Medical Research Council, 1991, *Ecologically Sustainable Development, The Health Perspective*, October, NH&MRC, Canberra.

Nierenberg, W.A., 1993, Science, policy, and international affairs: how wrong the great can be, *Environmental Conservation*, 20: 195–197.

Odum, E.P., 1959, *Fundamentals of Ecology* (2nd edn), W.B. Saunders, Philadelphia.

Odum, E.P., 1992, Great ideas in ecology for the 1990s, *BioScience*, 42: 542–545.

Odum, W.E., 1982, Environmental degradation and the tyranny of small decisions, *BioScience*, 32: 728–729.

Ophuls, W. and Boyan, A.S., Jr., 1992, *Ecology and the Politics of Scarcity*, W.H. Freeman, New York.

Parry-Jones, W.L., 1990, Natural landscape, psychological well-being and mental health, *Landscape Research*, 15(2): 7–11.

Richmond, M., 1993, Science and wealth creation, *Nature*, 362: 584.

Sahagian, D.L., Schwartz, F.W. and Jacobs, D.K., 1994, Direct anthropogenic contributions to sea level rise in the twentieth century, *Nature*, 367: 54–57.

Santana, C.E. and Jardel, P.E., 1994, Research for Conservation or Conservation for Research, *Conservation Biology*, 8: 6.

Sarokin, D. and Schulkin, J., 1992, Environmental economics and responsibility, *Environmental Conservation*, 19: 326–330.

Seligman, C., 1989, Environmental ethics, *Journal of Social Issues*, 45: 169–184.

Shaw, P., 1993, Warfare, national sovereignty, and the environment, *Environmental Conservation*, 29: 113–121.

Smil, V., 1993, *Global Ecology Environmental Change and Social Flexibility*, Routledge, London.

Soule, M.E., 1986, Conservation Biology and the 'Real World', in: Soule, M.E. (ed.), *Conservation Biology, The Science of Scarcity and Diversity*, Sinauer Associates, Sunderland, MA, pp. 1–12.

Soule, M.E., 1991, Conservation: tactics for a constant crisis, *Science*, 253: 744–750.

Sprugel, D.C., 1991, Disturbance, equilibrium, and environmental variability; what is 'natural' vegetation in a changing world? *Biological Conservation*, 58: 1–18.

Stern, P.C., 1993, A second environmental science: human–environment interactions, *Science*, 260: 1897–1899.

Thomas, V.G. and Kevan, P.G., 1993, Basic principles of agroecology and sustainable agriculture, *Journal of Agricultural and Environmental Ethics*, 6: 1–19.

TRSNAS, 1992, *Population Growth, Resource Consumption and a Sustainable World*, The Royal Society (London) and National Academy of Sciences (Washington DC) Joint Statement, Typescript, 4 pp.

United Nations, 1992, World human population now totals 5.5 thousand millions, *Environmental Conservation*, 19: 268.

Upretii, G., 1994, Environmental conservation and sustainable development require a new development approach, *Environmental Conservation*, 21: 18–29.

Vittachi, T., 1989, Demographics and socioeconomics: the people factor, in: Western, D. and Pearl, M.C. (eds), *Conservation for the Twenty-first Century*, Oxford University Press, New York, 26–28.

Walton, D.W., 1994, Australia, cultural landscapes and other Australians, in: Longmore, R. (ed.), *Biodiversity – Broadening the Debate*, 3, Australian Nature Conservation Agency, Canberra, 21–36.

Walton, D.W., Forbes, M.A. and Thackway, R.M., 1992, Biological diversity, environmental monitoring and the nature conservation estate in Australia: relationship to ecologically sustainable development, in: Longmore, R. (ed.), *Biodiversity – Broadening the Debate*, Australian National Parks and Wildlife Service, Canberra, 24–40.

Westing, A.H., 1993, Human rights and the environment, *Environmental Conservation*, 20: 99–100.

Westing, A.H., 1994, Population, desertification, and migration, *Environmental Conservation*, 21: 110–114.

Wilson, E.O., 1988, The current state of biological diversity, in: Wilson, E.O. (ed.), *Biodiversity*, National Academy Press, Washington, DC, 3–18.

4 Human health as an ecological problem

Napoleon Wolanski

Abstract

Health is not a clearly defined notion, although we all feel we know what it is. This vagueness derives from the subjective nature of health. Health means fitness to live, with a feeling of well-being, and all these elements – 'fitness', 'feeling' and 'well-being' – are subjective. While health escapes precise measurement, its description requires a definite point of reference which must be linked to the environment. Health reflects the relationship between an organism and its environment in both time and space. This chapter discusses four main inter-related problems: how to define health in terms of environmental criteria; how to measure health using positive indices (such as body build and physical fitness) in relation to those values from local populations, which have been moulded by the environment; limitations of negative indices of health (morbidity and mortality); and how the relationships between human populations and the environment change in the contemporary world, with the effects of modern civilizations and new pathological phenomena.

A conceptual overview

Health is not a clearly defined notion although we all feel we know what it is. This vagueness derives from the subjective nature of health. While health escapes precise measurement, its description requires a definite point of reference, which could be found in the environment.

Health could be understood as a 'certain' psycho-physical condition of humans determined by a 'proper' structure of the organism as well as a result of a 'dynamic balance' between the organism and its environment (homeorhesis).

Dynamic balance, or homeorhesis, is a developmental homeostasis over the time of ontogenetic development, related to the age of the subject and environmental conditions. It is changing (dynamic) homeostasis. The words 'certain' and 'proper' are indicative of how relative the evaluations may be.

Health is a vector, showing a range from good to bad health. This meaning of health could be applied, with slight modifications, to entire populations. Health and its symptoms vary substantially and dynamically with time and space. The health of an individual is not a static state over a lifetime. While each individual's health varies with time, the state called 'health' describes some typical properties acquired during phylogenesis, in relation to external conditions as a systemic element of nature. This has shaped *Homo sapiens recens*. The health of an individual depends on that individual's genes and stage of development, on the one hand, and the evolution of that individual's environment, on the other. In other words, the health of humans as a species is adequate to phylogenesis, while the health of an individual is adequate to its specific features.

It is true that humans can feel differently at various stages of their life. While internal organs wear out (structurally and functionally), this does not mean that as we get older our health declines. Health is a relative state at every stage of ontogenesis. The capability of human organs diminishes with age and so changes the relationship between the organism and its environment. Another point is that health undergoes some natural changes as the organism improves its organisation in youth (loss of entropy) while during the later stages of life the entropy increases.

However, despite all this, some individuals could be considered healthy because this would be assessed in relation to other individuals of the same age (or more accurately the same stage of development). Therefore health may 'diminish' through unfavourable internal changes of an organism, triggered by disturbances in organism–environment balance. As this occurs over time, it may give an impression that health indeed gets poorer with age, but this is not strictly true.

In a way health could be compared to physical beauty: it varies between ages and epochs. Health is relative in a similar way, but it can be defined at any one time with a set of criteria.

Similarly, the spatial aspect of health should not be overlooked, since health reflects the relationship between an organism and its environment including various geographical conditions as well as the human-made and social environments.

Individual traits of an organism interact with one another in a variety of ways. Some of these relationships remain stable throughout all, or the greater part of, ontogenesis; that is homeostasis. Other relationships between individual traits vary, as an organism is transformed through different stages of development; that is homeorhesis.

Such distinctions are sometimes used as a basis for distinguishing different phases of ontogenesis (Wolanski, 1972). Still other relations between traits of an organism change only temporarily depending on external conditions (e.g. vital capacity versus lung ventilation, haemoglobin concentration versus haematocrite index, blood morphology versus blood pressure). All such changes either complement or compensate one another in supporting an organism to reach some desired state.

Health can be conceptualised as an outcome of relations within an organism, and between that organism and its immediate surroundings. Thus a number of reference points could be distinguished which are not concurrent but form a cause–effect relation responsible for a given state. That state is reached in the process of development, which must be understood as a result of continuing self-adaptation.

The habitat, which is a set of environmental conditions, provides a test for the qualities of an organism and the process works also in the opposite direction. One could therefore say that it is a two-way dependence as regards both the shaper and the shaped. Usually, in the process of changing its environment, the organism itself undergoes changes. Hence, what we have here is not quite a passive–active relationship, but a self-regulating system. This fact alone provides the foundation for a thesis that health could be assessed only partly when the conditions of the environment are disregarded; going further, one could say that in some instances such evaluation is not possible at all.

'Health patterns', or more accurately 'morbidity patterns', are passed from parents to children. This not only occurs by genetic predispositions, but also by the transmission of cultural traditions and behavioural patterns in families, from generation to generation.

For the above reasons an evaluation of one's health requires information on the processes that have been going on before the examination stage and also requires some knowledge of the accompanying conditions. When this is not possible, reconstruction of the necessary information can be attempted by way of an intellectual exercise, at least for one of the sides, organism or environment. Clues are thus provided as to the condition of either the organism or the environment.

This problem is far from simple since health is a relative state. It articulates the equilibrium between the needs and aspirations of the organism on the one hand, and the ways and means of their fulfilment on the other. Health as the articulation of this equilibrium requires some internal order of the organism in relation to homeostatic, physiological, and psycho-physical harmony, and also regarding suitable external conditions.

What is unique about human beings is aspiration – a mental sense of the realisation of their plans. Because of the strength of people's socio-cultural backgrounds, this aspect of aspiration is no less important for humans than their structural biological needs. Therefore we must see the organism framed not only within the real conditions of the environment, but also by the background of the imagined needs which affect the human psyche and the shaping of humans' biological properties.

The appraisal of any phenomenon requires a criterion. Regarding human beings this criterion should and could be found in the environment. People's morphological properties, moulded by the environment, condition their needs in relation to their surroundings and make these needs relatively constant.

Health is a state of balance, the dynamic equilibrium between an organism and its environment, as defined by the organism itself. Thus one could say that it reflects a subjective point of view, a human perspective. Obviously a different approach, for example from the point of view of the world of plants or other animals or, more generally, an environmentally oriented view, would stand in contrast to this human approach. This 'environmental point of view' is very characteristic of ecology but it is by and large overlooked or even purposefully rejected in medicine. This is a strategic mistake since the shaping of environmental conditions also includes concern for human populations. Neglect of this point of view is responsible for the present ecological crisis. While it may seem that we are only on the threshold of this crisis, we are close to ecological catastrophe.

Environment

External environment usually denotes the natural and/or social surroundings of humans – a set of often inter-related elements with a direct bearing on humans. This understanding is, however, rather complex and therefore needs some comment regarding both its structural and its functional aspects which, in turn, brings us to the etymology of the word itself.

Environment is not synonymous with surroundings. The word 'environment' derives from the word centre and denotes an arrangement or system integrated by its centre. (In French *milieu* means centre, in Polish *srodowisko* comes from *srodek* meaning centre, in Spanish *medio-ambiente* also derives from the word *medio*, centre.) Of course, such a system could have a space of its own but its essence lies in the system of relations. It follows that the centre and its surroundings enter into a direct relationship; hence the environment of a given centre is formed only by the elements from the surroundings which remain in some direct relationship with the centre. The environment of an individual is his niche: his 'nest', which determines access to natural resources, use of technical environment, and biological and social health and fitness (contribution and satisfaction). The niche is a habitat and form of its exploitation, and contains all essential components of an environment that a subject or population needs for continued existence. In the same way we can tell that the proper niche of population and individual determines proper biological status of population and human being (health).

These related elements of the surroundings are not in all cases indispensable for a given object or centre – here we have in mind neutral, temporary or aggressive elements – but they are all necessary. In this context it is irrelevant whether they are objectively necessary, for example for chemical processes, or subjectively necessary for feeling good. The necessary elements include stabilising elements and destabilising ones. In ecology a set of surrounding elements directly related to an object is called a 'habitat'.

Of considerable importance is the amplitude of change of the environmental elements (e.g., temperature, composition of the air), which could be called 'factors'. Factors include also elements other than those of the environment (the second type being the external, 'exogenous factors'). The development factors also include the structures of the organism itself ('endogenous factors' both genetic and 'paragenetic'). Other factors are: physical activity, sleep, rest – related to lifestyle.

Of the four groups of factors mentioned above we discuss here only the exogenous factors.

An organism living in a particular environment is adapted to it, insofar as the range of factor change is concerned. Quite often the composition of factors is also relevant. In addition, the organism is adapted to any factor with which it has come into contact often enough and for a sufficient period of time to be shaped by it. Further, there can be factors which have stabilising effects at a given stage of development after the organism had become adapted to them. Hence, the development of the organism is a process of constant 'adjustments' (adaptive changes).

A factor whose amplitude exceeds the habitual amplitude for an organism causes changes whose effectiveness increases in line with the pace of ontogenesis in a given period. They may begin with a functional change followed by structural ones. This amplitude is called an 'ecological factor' or 'modifier'.

Modifiers can be natural or cultural. The former refer to bio-geographic factors, and the latter to social, social-economic and 'human-made factors'.

- The 'natural factors' include water resources, mineral resources, air composition, type and quality of living organisms, climate, configuration of the area, force of gravity and acceleration, vibration and sound, and electro-magnetic field.
- The 'cultural factors' include inter-familial relations and relations within the community.
- The 'family factors' are intellectual level (education background, type of schooling and upbringing), profession, earnings, size of family, the structure of the physical and emotional context of daily life at work, and recreation patterns at home.
- The elements that are external to the family (although usually interconnected) refer to ethnic group, size and character of the community, value systems, norms of co-existence, habits and customs. One type of modifier could be the impact of change in surroundings when moving house, even if the physical and cultural parameters remain unaltered.

An ecological modifier is, then, an extreme range of some environmental factor which causes a change in the organism. Only approximate values of that range are possible for a given area or population; the value is slightly different for each organism because of genetic properties and experiences. The modifier makes the organism 'start a game' with its environment for fear of experiencing an uncomfortable disturbance of the general equilibrium with the environment. The objective of the 'game' is to recapture balance. This means accommodation to the new amplitude or adaptation to the environmental factor itself.

It should be recalled here that all things in nature are inter-related and that there are feedbacks operating in the transmission of the stimuli. Thus a change in the environment would as a rule cause 'self-compensation' in other elements in the same habitat. Eventually, the organism is exposed to more than one modifier; indeed it is exposed to a whole system of altered environmental conditions. The same process operates within the organism itself: compensation changes occur and a new balance is attempted.

A fine-tuning of those two self-regulating systems is frequently much more complex than one might expect. An additional factor is the mental aspect of human adaptability.

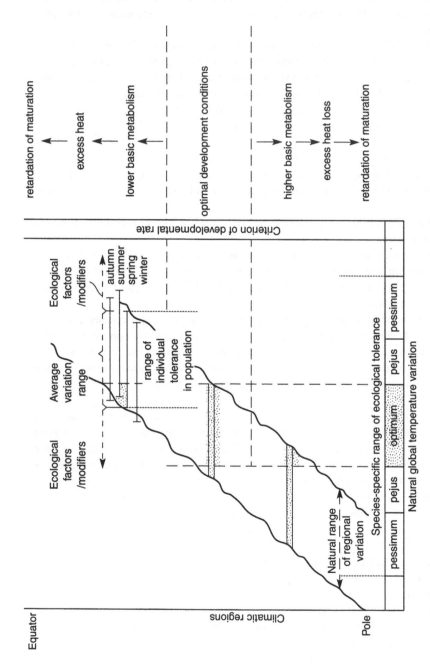

Figure 4.1 Zones of biological tolerance of species, in relation to climatic zones
Note
This illustrates the influence on the rate of development, and its connection with the human metabolism

The optimal range is one to which the organism becomes accustomed in the course of evolution. Therefore when we say optimal range we do not have in mind a conventional meaning, but rather the product of millions of years of phylogenesis. In other words, it is a 'genetic adaptation'. Conversely, the adaptational changes (adjustments) during ontogenesis are only non-hereditary modifications.

As far as typical features are concerned, for example skin colour, as a result of adaptation to the volume of sunshine, we can also distinguish strongly negative features ('pessimum') both with regard to the amplitude of individual factors and to the overall properties of the habitat (certain regions). Between the positive and negative limits there is a range called 'pejus'. It extends beyond the optimal values though without any immediate threat to life; it is a group of conditions with the greatest accumulation of pathology-generating modifiers (Figure 4.1).

Optimal conditions are measured by the economy of physiological processes; optimal conditions are most economic. Another distinguishing feature is the scarcity of negative stressors ('distressors'), or more generally 'des-adaptive modifiers'.

Other criteria applied here include fertility (reproductive fitness); however, considering such phenomena as health and sickness and their possible consequence such as death (selection in evolution) it would be more appropriate to apply the criterion of survival and well-being.

Organism

The basis of ecological regularity is that organisms are linked with their environment by their 'live needs', which are functions of existing morphological adaptations and as a result are stable. Organisms are equipped with safety mechanisms which protect against excessive or uncontrolled action from modifiers. Some of these mechanisms (mechanical, chemical) form permanent safety barriers. Others are activated when the toleration limits to external stimuli, the genetically determined 'norms of reaction', are exceeded. The relation of safety mechanisms to homeostasis is not yet quite clear.

Another indicator of the relationship between the organism and its environment is its sensitivity to environmental stimuli, its 'eco-sensitivity'. This represents the level of resistance to external stimuli. Eco-sensitivity is genetically determined and modifies conditioned reflexes. It is the prime detector of boundaries between the factors which determine tolerable values for the organism and the range of modifier operant in the ecological setting. Currently, it would seem that eco-sensitivity is greater in heterozygotic organisms than in homozygotic organisms. In humans, eco-sensitivity is greater in children aged between two and three years (after the period of maternal protection and breast feeding), during early pubescence, and in elderly people. In fact, the organism is most vulnerable in the phase of cell proliferation and organ development during embryonic growth, although the embryo is protected by the placenta.

Whether eco-sensitivity is sex-conditioned is debatable. There is no agreement on 'adaptability' and safety mechanisms in men and women. Generally women

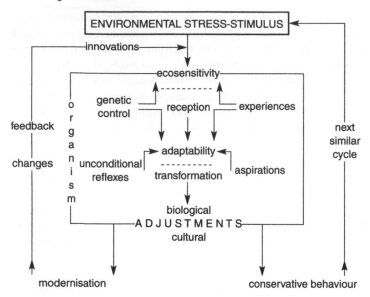

Figure 4.2 Model of eco-sensitivity and adaptability of organisms to environmental
stimuli

Note

An organism's reaction – the adaptive biological or cultural change – is always proportionate to the
strength and duration of the external stimulus. The longer the stimulus lasts and the stronger it is,
the greater the effect. The organism is characterised by its adaptability, which is a neuro-hormonal
feature that strengthens or neutralises the received stimulus. This depends on the organism's
genetically conditioned sensitivity, modified through phylogenetic experiences (unconditioned
reflexes) and ontogenetic experiences (habits, conditional reflexes) and aspirations.

tend to react faster and more strongly to stimuli but adaptive changes in women
are less penetrating than in men, possibly because of the need to protect the
foetus.

Until recently the organism–environment relations have been treated rather like
a 'black box'. Human ecology has been preoccupied with adaptability and various
environmental stimuli.

For several years now, more effort has been devoted to the development of
environmental physiology. Lately we have presented some hypotheses regarding
the relationship between eco-sensitivity and adaptability, which attempt to merge
those two into one called 'reactivity'.

Depending on individual eco-sensitivity, a stimulus is registered differently by
the organism. As mentioned earlier, eco-sensitivity is probably genetically determined and modifies conditioned reflexes at the level of individual experience of
the organism. A stimulus is thus initially processed by adaptability, before the
organism 'decides' on a reaction.

As for the genetic side of adaptability, we make reference to 'unconditioned
reflexes' (species experiences). However, the modifying factors are also the experiences gained during development (conditional reflexes, aspirations, ambitions),

Table 4.1 Mechanism for maintaining equilibrium between a population or an organism and the external environment

Environmental stimuli	Organism's response	Mechanism	Examples
Single short-term stimulus	Short-term defensive changes or mobilisation of organism to survival	Regulatory adjustment (reaction)	Defence reflexes, changes in pulse and breath frequency, immunological and emotional reactions
Longer-lasting repetitive stimuli of moderate strength	Morpho-functional reversible changes	Acclimatory adjustment	Acclimatisation, building of resistance, changes of behaviour, working hypertrophy of muscle tissue, increase in fat cell mass
Long-term strong stimuli	Morphological irreversible changes	Developmental adjustment (phenotypic plastic changes)	Changes in bone structure, increase of cell number, change of personality – consequences in ontogenesis
Very rare stimuli – decisive for survival	Inability to survive, death (or infertility)	Adaptation (genetic)	Selection, death of unadapted individuals (or infertility), survival of adapted individuals – consequences in phylogenesis

so characteristic for humans (Figure 4.2). 'Aspirations', if considered as motivation, may considerably influence the way the organism chooses to adapt to a situation but in an indirect way, i.e., through emotional experience.

A single and relatively short and low-impact stimulus causes a certain reaction if it is stronger than the level the organism can take (Table 4.1 and Figure 4.3). These are temporary and reversible changes devised to prevent a more penetrating change called 'regulatory adjustments' (reaction). A stronger, longer and repetitive stimulus usually gives a functional, and sometimes even morphological change, which is more lasting but still reversible. This is called 'acclimatory adjustment', and examples are the process of getting used to new climate conditions or the over-growth of tissue under strain.

Strong and prolonged stimuli – which, however, do not destroy vital organic structures – give 'developmental adjustments' (also called 'phenotypic plastic changes'). These are irreversible, morphological changes going on during ontogenesis. Individual differences in development, along with genotype differences, comprise the effects of adaptations. Such variations give rise to a differentiation in the needs of the organism.

Humans have great adaptation capabilities in the field of biological change

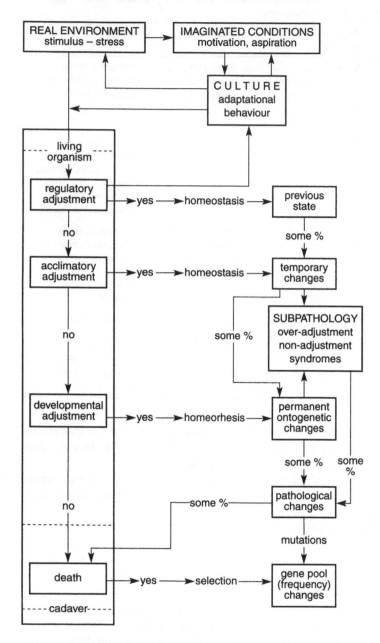

Figure 4.3 Model of adaptational changes of an organism in response to environmental stress (bio-adjustment)

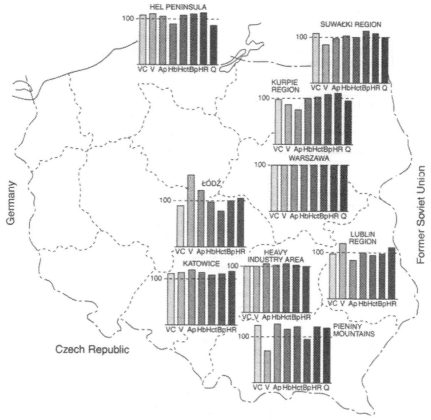

Figure 4.4 Adaptive changes of cardio-respiratory functions of humans, measured in
various climatic conditions in industrial and urban settings in Poland

Note
VC = vital capacity of lungs
V(VE) = lung ventilation at rest per minute
Ap = interval between breaths
Hb = haemoglobin concentration in blood
Hct = blood cells in percentages of blood volume
BP = arterial pressure of blood
HR = heart rate at rest
QR = minute heart volume at rest (cardiac output)
Source: Pyzuk and Wolanski (1972); Koziol (1989)

and also have such capabilities responsible for culture-generated ways of fore-
stalling such changes. Biological changes, however, may not necessarily be
advantageous even though they could be adequate environment-wise. The two
almost opposite kinds of adaptation may be distinguished as 'over-adaptation' and
'non-adaptation'.

Over-adaptation occurs when excessive change in the organism drains our
adaptability reserves. While such disproportionate change is usually ineffective and
costly, for instance in terms of high energy consumption, a depletion of reserves

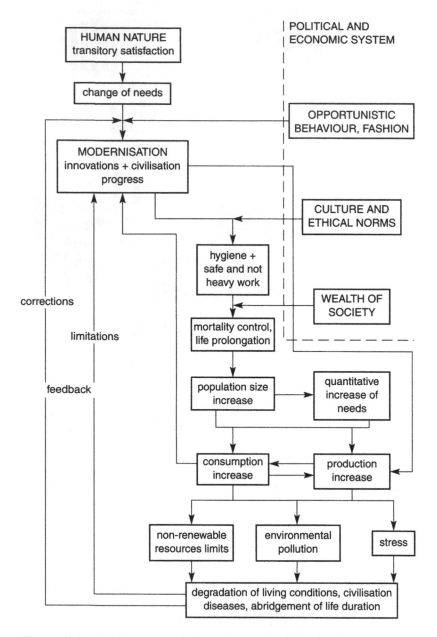

Figure 4.5 Model of adaptive behaviour in modernisation: natural and socio-cultural
environments

is a handicap when disturbances keep on repeating. Let us consider the adaptation of breathing and circulation reactions in various climatic conditions. At sea-level, we can observe enhanced ventilation of the lungs, higher blood pressure and lower haemoglobin index and lower cardiac output. At moderate altitude, lung capacity enlarges, haemoglobin level increases, as does cardiac output, while lung ventilation drops and blood pressure falls. Such behaviour illustrates the compensation phenomenon where one function makes up for another involved in the same process, in this case tissue oxygenation. Severe pollution can intensify all those functions and structures. While it is possible for an organism to enlarge the heart capacity and haemoglobin level at sea-level, or to improve ventilation and blood pressure at altitude, such options do not exist or are dramatically limited in industrial areas (Figure 4.4).

Non-adaptation means an inability to adapt to new conditions, inadequate changes or, in extreme cases, a general (rather than total) loss of adaptability. As mentioned earlier, the organism's final response to external stimuli represents various kinds and degrees of adaptive change. They may be either biological changes of the organism, or cultural changes in its behaviour and efforts to modify the impact of the environment. The latter, understood as rational action, is characteristic of humans only, and constitutes a complete cycle of human bio-cultural relationship with the environment, illustrated as a model in Figure 4.5. The model depicts a comprehensive approach to human adaptation mechanisms. It could prove useful in analysing the causes of disturbances in relations between the environment and humans and could assist social diagnosis for the restoration of equilibrium.

As far as the operation of receptive mechanisms, stimuli and adaptation is concerned, health might be discussed in terms of the environment. The health problems are part of the problems of the human niche: health is not only a biological problem, it is also a cultural one. The human niche is not only a biological but also a socio-cultural net of human and environmental inter-relations.

Organism–environment relation: a criterion of human health

Health, from a total environment perspective, means on the one hand adequacy of the construction and function of organisms, and on the other the conditions in which organisms are formed and live. Thus health is a state of 'dynamic balancing equilibrium' of the organism in relation to environmental conditions.

However, this criterion should not be applied rigidly in view of the tremendous flexibility of the human organism. Such adequacy means the satisfaction of both biological needs rooted in the evolution of the species, and of those aspirations arising from the cultural foundations of our society. Health could thus be expressed for an individual human being as follows:

$$\frac{\text{satisfaction of daily needs and aspirations}}{\text{biological needs and aspirations past and present}}$$

In fact health should be measured by the functioning of the organism in the light of objective criteria and individual assessment vis-à-vis a statement on how it should function in given conditions (needs).

One could conclude from the above that a typical individual formed in an environment (habitat) which might be considered optimal becomes a biological system of reference, i.e., a positive gauge of health. Given the difficulties involved in such measurement one has to take into consideration also the negative measurements of health, i.e., the assessment of ill-health or sickness. The ecological concept of good health concerns harmony and compatibility of the organism and the environment. The concept of a lack of health is based on inconsistencies between the organism and the conditions of the environment, brought about by unfulfilled needs and aspirations or else caused by undesirable actions in respect of species and individual criteria.

Such environmental changes (evo-deviations), which disturb the relations between the organism and its living conditions (causing maladjustments) should be called negative. They include a total change of habitat: new natural surroundings, different social and living conditions, different ethical and cultural norms, and human-made conditions. The destruction of human–habitat interconnections leads to a loss of health. A majority of social and cultural factors influence the biology of people through psychological experience.

Changes in the natural surroundings, new social and hygiene conditions and different economic status have a direct impact through climate, food, sources of infection, and so on. Further, the human-made surroundings of humans have their share in the change of physical and chemical conditions as well as in the shaping of mental feelings: beauty, harmony, sound or colour patterns, and suchlike.

Any modification of habitats to which our species is adapted and to which individual human being are adjusted, is potentially unfavourable. Yet, humans inhabit many places on the Earth where biological harmony with the surroundings is not possible. Harmony can be achieved through protection of cultural barriers against adaptation selection. Additionally the issue gets more complicated by humans' cultural determinants of health, i.e., through common views towards health. Individual mental and physical comfort is exposed to strong social judgement: both positive, taking deviations for normality, and negative, taking a normal condition for incorrect behaviour. 'Normal' is defined here as average, proper and favourable environmental conditions. 'Correct' is understood to be adequate to an adopted criterion (Wolanski, 1961a).

The causes of disease associated with environment are as follows:

- difficulties with biological adaptation;
- excessive stress working through the nervous system on the psycho-physical condition;
- growth of unfavourable physical and chemical stimuli, including carcinogenic and allergenic, particularly notable in connection with prolonged life-span. An important problem is environment-induced mutations.

Genetic determinants and environment

A number of mechanisms of genetic-ecological interaction, as one element in the adaptation of an organism to its environment, have recently been highlighted. We have in mind the so-called 'response norm', or norm of reaction, described elsewhere in categories proper for human ecology. The term 'reaction norm' as used until now is not broad enough to describe the reality, because the possible reactions are not limited to those of a defensive kind. Here are a few examples:

- With an increased variety and quantity of food, the probability that children's appearance will resemble that of their parents diminishes, probably as a consequence of a realisation of various alternative ways of development. This effect could prove unfavourable when conditions in habitats become worse, e.g., pollution of air and water, incorrect nutrition, mutant factors, and so on.
- Migration and disruption of isolates increase the number of 'heterozygotic' individuals in gene loci, thus promoting 'heterogeneous' populations. As a consequence, populations become more vulnerable to environmental factors. Under favourable conditions, this means a better utilisation of people's development potential because of greater sensitivity to external stimuli. In unfavourable conditions, this may favour unwanted changes.
- A certain optimal value of the genetic differentiation – heterozygosity – could be distinguished. This may also concern the so-called species similarity module (Wolanski and Siniarska, 1984a). When the homozygosity is too high ('homosis', 'inbreeding depression') or heterozygosity is too high ('heterosis'), fertility drops. Losses are greater and the offspring do not develop well. So the optimal value as it appears should correspond to the most favourable reactions of the organism to stimuli from the environment. This problem could be related to the concept of 'response norm', i.e., genetically pre-programmed limits to adaptive change. It could also be associated with the theory of a limited-direction development (Wolanski, 1971).
- Not all properties of an individual are equally susceptible to externally stimulated changes. There is something like a positive and negative 'responsiveness': propensity or capacity to change. Features which develop mainly during the post-natal period betray greater vulnerability to external changes; therefore they would eventually have to undergo greater adaptation change (Wolanski, 1975b; Wolanski and Antoszewska, 1990).

It is important from the point of view of the improvement of health (the preservation of good health) that individuals or traits more vulnerable to environmental stimuli are at the same time more likely to face disturbances of development, though usually they are also better prepared to regain the balance in favourable circumstances.

Today we could say that there is some evidence for the existence of protective mechanisms against excessive vulnerability or susceptibility to change specific for every individual, for example, assortative mating. It seems quite likely that what we are dealing with are even larger-scale ecological mechanisms operating within whole ecosystems. Therefore a lot of effort should now go into determining how far humans are altering the regularities governing their health and their very existence.

Positive indices of health

In keeping with the WHO definition, there are negative and positive health indices. The negative indices on which all statistics of the health service are based present few difficulties. As regards the positive indices of health, at least three elements have to be considered: measurements and tests recording the subject's state of health, methods evaluating the degree to which a given subject remain within the accepted normal range, and standard values reflecting 'biological reference systems'.

In this author's earlier work, consideration of biological reference systems (Wolanski, 1959, 1968, 1974) highlighted the need for discussion of the environmental criterion of health. This criterion was favourably accepted, and is now fairly widely applied, particularly by paediatricians and nutrition workers.

Measurements and tests used for recording the state of health by positive indices vary widely. For practical purposes their range must be minimal in screening tests (prophylactic mass examinations), whereas it has to be optimal in clinical tests, and must be unlimited in research work. Funding constraints usually limit studies of this kind to a few selected measurements of morphological and physiological traits, and to psychomotor tests.

Methods of analysis of growth and development of children against the scale of the population, based on mean values, involve three aspects:

- time-dependent changes in a given population, measured against the scale of a region, in particular in villages, or of a (small) country;
- differences between various urban and rural populations of a given country;
- comparison of the results for the population of a given country with those from other countries.

Indices of the status of organism and environment

Although it may be obvious, the point has to be made that both an analysis of factors influencing the state of health and evaluation of the state of health itself are usually based on indirect rather than direct indices.

Income per family member is an important element characterising the living standard of the family environment, because it is strongly correlated with other elements of living conditions. Whereas the amount of money is not a factor influencing the state of health, and while money does not directly affect health, it determines to a great extent the possibilities of realisation of human needs. Income is strongly correlated with many intra-familial factors exerting an effect on the state of health, such as education which determines for example the awareness of intra-psychic and nutritional needs, and with the occupation which greatly influences lifestyle, geographical mobility, and so on.

Therefore, income per family member is an index which better than any other characterises the complex of socio-economic factors in the family. However, increase in a family's economic status will not result in improvement in the child's health without a concomitant increase in the parents' level of education. Smaller families or larger houses/apartments do not necessarily result in improved child development (Siniarska, 1994, 1996). Education and occupation combined give an accurate prognostication of the state of health, although income is even more accurate in this respect. (Wrebiakowski *et al.*, 1982). It remains unknown to what degree the above selection of indices, suitable for Poland in the 1970s, is appropriate for other societies.

The stature of children and young people is the most frequently used index of a positive state of health, in spite of reservations concerning the validity of such considerations. For example, newborns with large birth-weight can experience significant problems. Tallness in a subject does not necessarily testify to his normal development; his stature has to be evaluated in relation to that of his parents. Stature is the result of a complex of developmental processes. Subjects brought up under good living conditions are tall, and those from adverse living conditions (malnourished, often ill, etc.) are short. Stature therefore indirectly indicates that the subject was brought up under specific environmental conditions. If the stature of an individual exceeds that of his coeval, and if the difference is not the consequence of his parents' being taller than other parents in this population, then it can be accepted that his stature results from favourable living conditions. Since the living conditions determine the overall phenotype of an individual, it can obviously be assumed that other traits of the organism are also positive and satisfactory. Thus it is expected that tallness on a population scale testifies to generally satisfactory functioning of the organism, i.e. – indirectly – to good health. Whereas in this case the reasoning involves well-grounded assumptions, it is based only on indirect relationships, and therefore stature (like any other single somatic index) used as an indicator of the state of health must be regarded as giving only an approximate and relative picture. Stature is a more reliable index for evaluation of whole populations than of individual subjects. On a population scale, the results of various effects compensate each other, thus affording a more correct evaluation, owing to the acceptance of the criterion of ecological conditions, which is usually characterised by Gaussian distribution of the different environmental elements. The respective extremes in the distribution of living conditions call for analogous extremes in the distribution of the investigated traits of a subject. Of course, this relates to the quantitative traits, also called traits with continuous distribution. The values of traits occurring between the extremes and the values calculated as averages (involving the same subjects, but with a different procedure for establishment of the 'normal range') would point to an average state of health of individuals, corresponding to average living conditions in a given population.

Table 4.2 Recommended selection of variable for the monitoring of human
           biological status

Somatic traits	Physiological traits	Psychomotor traits
	Minimal programme	
Stature [body size]	*Forced Expiration*	*Movement accuracy*
Kaup index [body build]	*Volume 1 sec.*	[proprioceptive feeling]
(weight:height²)	{FEV1sec}	*Grip strength* [muscular
Cephalic index [head shape]	*Haemoglobin*	strength]
(head breadth to length)	*concentration* {Hb}	*Standing long jump*
		[explosive power]
	Optimal programme *(as above plus)*	
Subcutaneous fat thickness	*Diastolic blood*	*Palant ball throw*
(on arm at triceps brachii)	*pressure* {DBP}	[explosive power]
Shoulder–hip index	*Working heart rate*	*Shuttle run* [agility]
[trunk shape] (shoulder	*at 5′* submaximal	*Sit-ups* (or Burpee test)
to hip breadths)	work load {HR 5′}	[dynamic persistence]
Chest circumference [ch. size]		*Kraus–Weber – psoas test*
Face breadth and height		(or hanging by arms)
[face size]		[static persistence]
Nose length [cartilag. tissue]		*Balance testing*
Frontal breadth		[coordination]
[prosencephalone size?]		*Spine flexibility*

Source: Wolanski, 1994a
Notes
Terms in braces are abbreviations of variables.
Terms in square brackets indicate general characterisation of traits.
Terms in parentheses are measurable traits needed for calculation of variables or alternative traits.

Studies have shown that among somatic 'polygenic traits' for the best positive
indices of the biological status of human populations it is possible to select several traits, which are correlated with others (Table 4.2, Wolanski, 1994a). Among
somatic traits, three were selected for a minimum programme: stature, Kaup
index (BMI), and cephalic index, based on four measurable traits: body height
(B-v) and weight, head length (g-op) and breadth (eu-eu). Among physiological traits, two were selected: forced respiratory volume per second and
haemoglobin concentration. Among motor traits, three were selected for the
minimum programme: distant movement accuracy (proprioceptive feeling), grip
strength (static muscular strength), and standing long jump (explosive power of
lower extremity).

The traits selected for a slightly extended, optimal programme included in
addition six somatic traits: subcutaneous fat thickness, shoulder–hip index, chest
size, and face, nose and frontal size, based on eight measurable traits: subcutaneous fat thickness on arm at triceps brachii, shoulder breadth (a-a), hip breadth
(ic-ic), chest circumference at rest on xiphiale (xi) level, face breadth (zy-zy), face
height (n-gn), nose length (n-sn), and frontal breadth (ft-ft). Important additional traits for physiological studies consist of diastolic blood pressure (DBP) and

working heart rate (HR) at 5' of submaximal work load (Koziol, 1989; Koziol-Kolodziejska, 1992). Important additional motor traits in the optimal programme are distance of soft ball throw (explosive power of upper extremity), duration of shuttle run (agility), dynamic persistence (number of sit-ups per minute, and/or Burpee squat-thrust test), static persistence (Kraus-Weber test for psoas muscle and/or hanging on arms flexed position), balancing ability (time spent standing on one leg), and spine flexibility.

Biological reference systems

Indices of this kind, being satisfactory under conditions pertaining in economically developing countries, lose their importance in populations of highly industrialised areas with heavy environmental pollution (Wolanski and Siniarska, 1984b) and in countries with a high living standard and only slight social stratification (Lindgren, 1976). In greatly industrialised areas, more complex evaluation methods are indispensable, as will be discussed below. It is noteworthy that very often some single traits (in particular, physiological and biochemical ones), while seemingly giving a very accurate picture of the biological status with respect to, e.g., the blood haemoglobin level, are in fact very unreliable indices. For example, it is well known that in subjects of the same sex, age and nutrition level, the haemoglobin (Hb) concentration is greater in mountainous regions than in those bordering the sea.

Thus, upon assumption of the 'mountain standard', 'coastal children' would in a high percentage of cases appear to be anaemic; yet, this is not necessarily the case, if other traits of the organism, participating in oxygen transport to tissues, compensate for this feature.

It is found that children inhabiting foothills or low mountains usually have – apart from high Hb concentration – low blood pressure, as well as high vital capacity and low ventilation. On the other hand, coastal inhabitants have – apart from low Hb concentration – high blood pressure, as well as medium vital capacity and high ventilation. Therefore, the same effect is achieved in another way. In addition, the population of greatly industrialised areas, with high dust pollution, exhibits high values of all these traits. Therefore, each of these symptoms – taken separately – would evaluate positively the state of health of the population of the heavily industrialised area. However, the global increase in the value of traits participating in oxygen transport has to be interpreted as over-adaptation which is a sub-pathological symptom. In the presence of these compensatory changes (mountains and coastal regions), in case of a disturbance in the environmental conditions, the organism may temporarily bring about an increase in the low values of some traits. Under conditions of a greatly industrialised city, these possibilities are exhausted (compare Figure 4.4).

The three measures generally assumed to be replaceable: hematocrite index (Hct), Hb concentration and erythrocyte count, were found to be equally important for the diagnosis of the effect of various disturbances induced by environmental factors (especially environmental pollution and nutritional disturbances).

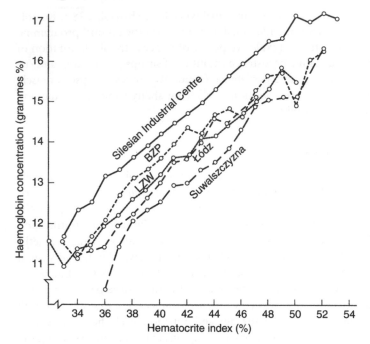

Figure 4.6 Haemoglobin concentration plotted against hematocrite index in inhabitants
of specified locations in Poland

Notes

Measurement is by direct methods

Areas covered (with abbreviations in parentheses) are: villages of agricultural Suwalszczyzna, the
textile-manufacturing city of Lodz, villages of the Lublin Coal Basin (LZW) experiencing the first
phase of industrialisation, towns of the Belchatow Industrial Region (BZP) also experiencing the first
phase of industrialisation, and towns of the Dabrowa Coal Basin (Silesian IC) in a heavily polluted
area of steel plants, etc.

Persons in the sample are in the age range 2–80 years

In particular, the Hct index and Hb concentration can reveal different changes
caused not only by extreme conditions but also by damage to the environment
inhabited by present-day Polish populations (Koziol, 1983). These relationships
are different for various age groups (Figure 4.6).

Local regional standards

If the environmental criterion of the biological status is applied, an analysis of trait
distribution is the basic reference system. However, for this criterion to be true,
evaluations of individuals have to be referred to the standard values of their own
population or of an analogous one. For a long time, stress was put on regional
standards regarded as the true biological reference systems (Wolanski, 1959,
1965, 1968, 1974).

Previously this view was controversial, whereas at present it is univocally
accepted in many countries (Peters *et al.*, 1982). The point is that the average

values of the population in which a given subject was brought up and lives are the only true reference system, as being determined by a definite specific complex of living, climatic, and social conditions.

Distribution of traits

Evaluation of the biological status of a population can be based on measures of the central tendency (e.g. arithmetic mean, if the distribution is normal) and on measures of dispersion (e.g. standard deviation, if the distribution is normal). In this case the reference system must consist of another 'standard' population, with the highest living standard on the scale of the country or a group of countries with similar socio-economic systems, geographic conditions and gene pools. However, for evaluation of an individual, only his own population has to be used.

This recommendation is insufficient. The reference system must comprise a population sample which has good living conditions and is selected at random, with the elimination of disabled subjects and of the genetically loaded or chronically ill. Only such material can be the basis of construction of, e.g., percentile grids (other methods are less susceptible to disturbances resulting from an uncorrected percentile system – Wolanski, 1961b). In the light of the theory of limited direction of development (Wolanski, 1971), it is essential to apply a division of distribution of polygenic trait variation, which permits observation of changes (particularly in children and youth) within relatively narrow channels. The course of development within a given channel is genetically determined, whereas 'jumping' from channel to channel testifies to developmental disturbances.

The natural range of normal variation consists of a standard deviation in both directions from the arithmetic mean (x ± 1 SD). In other words, it corresponds to the sites of 'inflexion' of the curve of normal distribution. Within these limits (intervals) 68.26 per cent of all cases are contained. Since the distribution (particularly during the development) is seldom normal, i.e. symmetrical, a percentile system is adopted. Therefore, the limits of the 'normal variation' amount to 70 per cent, rounding off the value above. Since these 70 per cent represent too wide a central segment, the 'wide normal variation' is divided into narrower segments: the central 30 per cent (from the 35th to the 65th percentile) and 20 per cent on either side of this 'narrow normal variation'.

This affords the following division:

wide   n o r m a l   v a r i a t i o n

percentiles   15 — 35 — 50 — 65 — 85

percentile content   20% & 30% &   20%

                      & narrow normal &
                      & variation &
                      + - - - - - - - - - +

Using this division, 15 per cent of cases remain outside the 'wide normal variation' on each side. This part of the population must be analysed more closely. From the standpoint of the degree of deviation from normal, these cases are not univocally abnormal. For instance, if stature is analysed, these cases may comprise children of a particularly tall or particularly short parent, whose stature is determined genetically. However, the group of tall cases may also include those with endocrine disturbances, and the group of short cases may include malnourished subjects, with metabolic and endocrine disturbances etc.

In this connection, the final 15 per cent are divided into a segment containing 10 per cent of cases (which ought to be analysed more closely) and the extreme 5 per cent which in fact are regarded as at least sub-pathological. The 10 per cent of cases ought to be subjected to extensive medical examination and to genetic analysis of parental traits, as well as to evaluation of living conditions. The remaining 5 per cent must be examined more closely (e.g. using morphograms and Decourt-Doumic profiles), to detect the possible disturbances in the inter-tissue composition, inter-extremities proportions, and so on. Thus established, the division assumes the following form:

```
                & wide    normal   variation &
                &    & narrow   norm &    &
percentiles — 5 —— 15 ————— 35 ————————— 65 ————— 85 —— 95 —
percentile  5%   10%       20%         30%        20%     10%   5%
content     pathol. risk                                 subpathol.?
            subpathol.?                                  pathol. risk
```

Of course, although the above system is based on a formal division, it is related to some regularities of Gaussian distribution (the wide norm is close to ± 1 SD), so it is grounded in the environmental criteria discussed above.

Other divisions used at present are increasingly criticised (Morabito *et al.*, 1982). The system 3–10–25–(50)-75–90–97 does not reflect the natural system of distribution within the population, and the range of the external 3 per cent is too narrow for the 'risk factor' (Nicoletti and Pelissero, 1979). Even by its author, this system is now recognised as erroneous (Tanner, 1981).

Another example related to positive indices of health may be evaluated on the basis of cross-cultural family studies (Wolanski, 1996). Optimal conditions for good biological status of children in Polish families (high level of education, high income, and small family – Wolanski *et al.*, 1988; Wolanski and Zaremba, 1996) are not optimal in Japan (Wolanski *et al.*, 1994a, 1994b). Under conditions of the European civilisation and low economic level of Poland, the above-average income typically signifies merely the satisfaction of basic needs.

Better education of parents influences only their ability to utilise their modest income to improve the health of their children. In the economic situation prevailing in Poland, a small family means fewer people to feed and clothe on a limited income. The same is true in Mexico (at least in Yucatan). This correlation does not hold in Japan or in Korea. Unlike in Poland and Mexico, a decrease in income in Japan and Korea presumably does not go beyond the social minimum. Thus, differences in conditions promoting the development of offspring are difficult for representatives of a distinct culture and economy to comprehend. In Japan, child development is optimal in large (probably traditional) families even if levels of education and income are below the average for the country.

In western civilisation, the family is generally based on the 'individual productive effort of man' (typically the father of the family), whereas in the Far East the counterpart is the 'family as a productive unit' (Leonetti, 1976) or (as in China) the family is a unit both of production and of consumption (Wang *et al.*, 1996). Ecological and cultural transmission of information between generations, even similar in any society, are different in contents, which effects various biological status of each population.

Differences between civilisations cannot be compared with differences between species, but they must not be neglected, as they represent distinct features making mutual understanding difficult. To a great extent, they influence the biological (health) status of local populations.

Negative indices of health and cultural maladjustments

It is a paradox that in some cases the evidence of illness increases with the number of physicians in a given area or population. This is an effect of more frequent visits to clinics, not worse living conditions or inferior health care. Because of such effects, negative indices of health are strongly criticised. The most objective negative index is mortality, but a percentage of mortality describes the status of a population, not of individuals.

Consequences of changes in the mode of life and in the environment are most readily seen at the beginning of the life cycle. Hence, infant mortality is considered as an index of the level of health service and economic condition of a country. It seems, however, that it is also an important index of social tension and cultural values. Infant mortality in Poland dropped over the postwar period from a mean of about 120 to 16 per 1,000 live born.

Fluctuations in this index in postwar Poland are closely related to fluctuations in the economy, as indicated by the analysis of changes in annual national income (Wolanski, 1991). Any acute change in the economy was combined with increased infant mortality, including an increase in the difference in infant mortality between towns and villages to the disadvantage of villages (Figure 4.7). Such periods were also times of increased social tension.

All these fluctuations, however, did not essentially change the position of Poland among European countries (Figure 4.8). It seems that this permanent position is primarily due to the health culture (diet, mode of life, standards of hygiene, stresses, and so on). Culture in this sense is as conservative as genes are biologically. The position of Poland in terms of health does not seem to be determined mainly by its economic ranking, and even less so by the organisation and development of the health service, especially with respect to infant mortality.

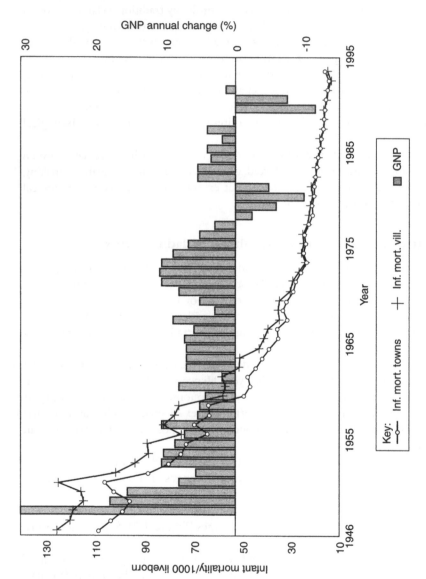

Figure 4.7 Infant mortality per 1,000 live births in Poland, 1946–1995, compared with annual percentage change in GNP

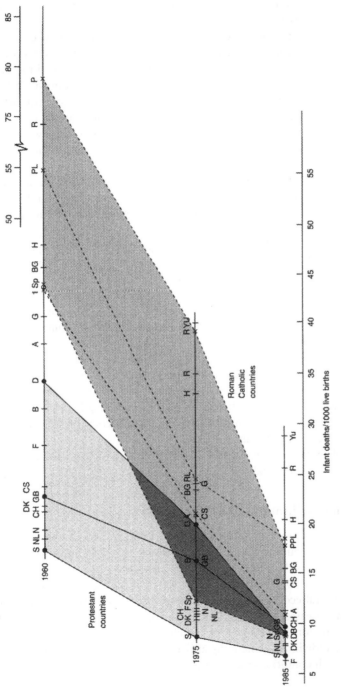

Figure 4.8 Infant mortality rates in selected European countries, 1960, 1975 and 1985.

Note

There are marked changes in four Roman Catholic countries (Portugal = P, Poland = PL, Spain = Sp and Italy = I), and in three typical Protestant countries (Germany = D, Great Britain = GB, and Sweden = S). During this period Italy moved closer to the rate typical of the Protestant countries, and Spain crossed the boundary from the Catholic to the Protestant group.

Most probably culture (health and hygiene practices) is responsible for the fact that Poland ranks between fourth and seventh among the countries with the highest infant mortality in Europe.

The poverty advocated by some ethical systems promotes neither economic development nor the progress of civilisation. Cultures that adhered to asceticism and the virtue of poverty are still characterised by poorer sanitary conditions, and have higher infant mortality (e.g. in such strongholds of the Roman Catholic Church as Poland, Italy, Spain and Portugal) than the Protestant countries which reformed their ethical and cultural systems. Making work and limited consumption a virtue – accepting the work ethic – countries such as Germany, Sweden and England made economic progress, raised the level of education, reduced infant mortality, and extended the life span of their population.

Among the people arriving in new industrial centres and towns in Siberia in recent decades, the incidence of disease has been 10 times that found among people inhabiting these areas for several generations, whose lifestyle is in harmony with conditions there (Alexeeva, 1986).

Early in this century, the Jews in Poland had higher incomes, better apartments, and higher standards of living in general, but their culture did not enhance their health because of the 'rituals imposed by culture' (Miklaszewski, 1912). In Poland high reproductive losses were recorded in the new harbour of Gdynia, constructed in the 1930s, and repopulated in the 1950s when the other port city, Szczecin, was the subject of a territorial dispute with Germany. Infant mortality in Gdynia was among the highest in the country over the period when the rate of immigration from other, mostly rural, areas was high (Wolanski, 1991). Rural cultural patterns revealed a biological inadequacy of these families to city life.

A culture developed under one set of climatic and social conditions, when indiscriminately transferred to a different ecosystem, including its socio-economic system, and combined with the maintenance of ethnic barriers between migrants and residents, leads to cultural inadequacy.

In ecological terms, a niche formed under given conditions cannot function under different environmental conditions.

Contemporary civilisation-affected diseases

Civilisation-affected diseases are protracted, long-lasting processes preceded by an asymptomatic phase during which the functioning of an organism is progressively impaired; in this they differ from invasive diseases.

The analyses made so far imply that the essence of civilisation diseases is their multi-factor, cultural-ecological etiopathogenesis, as independent of genetic inclinations to some diseases, environmental factors seem to be of major importance here. Such inclinations are releases or largely reinforced by different environmental factors. Individuals genetically afflicted with an inclination to a specific disease under such conditions incur a high risk of morbidity and are unlikely to recover.

A large group of diseases is closely related to the development of civilisation, especially when ecological factors are involved. The detection of causes in socio-cultural processes (mode of life) and in the effects of anthropo-pressure (pollution and degradation of the natural environment) is difficult. Factors enhancing civilisation diseases are typically unidentifiable, acting indirectly and in a very complex way, and often difficult to measure.

In a very general way, pathogenic factors can be classified as exogenous or endogenous. Although human organisms are autonomic units, they are in a state of dynamic balance with the environment (homeorhesis). Like every system, they exchange matter, energy and information with their surroundings. In this sense, each stimulus – exogenous or endogenous – changes the existing homeostasis. After the reception and recognition of information and its transformation into decision, the existing feedbacks release a response of the organism. The organism responds to a given situation individually, in relation to its own genotype and experiences, though typically this is an unconscious response. Responses to familiar factors are reflexive (involuntary, that is to say automatic) and genetically determined. A new, unknown factor requires time before the organism, usually using the trial and error method, will develop an appropriate response, that is, the range of defensive actions. All this seems to be relatively simple in this general formulation, when we consider pathogenic factors known to the organism, that is, those to which it was exposed earlier as an individual or as a species.

In just the same way open information stimulates the organism to a direct response of the type known as the stimulus-reaction effect. Most organisms, products of millions of years of evolution, function according to this rule. Mechanisms of struggling with natural pathogenic factors have been developed during phylogenesis of the species. This is a continuous process, as the evolution of the environment, including pathogenic micro-organisms, is in progress.

There are, however, many morbidity factors of a different nature, especially in socio-cultural environments, which are not explicit to an organism in the sense mentioned above. These are new factors, to which human populations were not exposed in earlier times. They have a character of sub-threshold stress in the sense that they do not stimulate at all, or they stimulate only to a small degree, the informative activity of an organism. They are acting for a long time, and the changes they cause are cumulative, whereas the organism remains passive to these changes for a long time. When these accumulated changes begin to disturb the functioning of the organism in a perceptible way, the organism shows signs of being surprised, and usually this response is already delayed, and ineffective for many reasons.

The response of an organism is characterised by a specific 'evolutionary wisdom'. Typically, this leads to the elimination of the cause of disturbance. In the situation described above, it is already impossible to eliminate the cause, and thus the organism responds in the only possible way. There are different adaptive changes (comp. Figure 4.3), depending on the duration and intensity of the stimulus, and also on the sensitivity to environmental factors (eco-sensitivity) and adaptability. In the cases described here such defensive response is often ineffective and, consequently, different symptoms of morbidity units start to develop. In this situation, these desperate responses of the organism lead to further disturbances of important physiological functions. Many examples can be given here, such as impairment of the pulmonary circulation or adaptive enlargement of the heart ventricles.

There are some environmental changes and stimuli to which organisms cannot adapt. For example, human organisms cannot adapt to lead or mercury, or in general to the group of elements competitively replacing nutrients, and being irremovable from the organism. Even if some adaptive changes take place, they are due to pathological rather than physiological reactions. This should be stated clearly because many people dismiss the seriousness of the problem of threats to the environment by arguing that organisms have almost unlimited capacity for adaptation. Indeed, the adaptability of living creatures is fascinating, but it has barriers determined by genetic norms of reaction: even if other sets of genes are activated, the effect will be only to shift the limits of tolerance. The essence of the matter here is the stage of ontogenetic development in which a stimulus is acting, its duration, and also the type and intensity of the stimulus. The consequences of such an apparent adaptation can lead to death of the organism or to degradation (degeneration) of the species in the future. Therefore, it is not a matter of chance that diseases of civilisation are also called adaptational diseases.

In the case of diseases of civilisation we deal with a complex mixture of many factors both exogenous and endogenous, natural and socio-cultural. The artificial environment created by humans changes the biosphere, producing substances unknown to living organisms, and wastes, as products of intensive production and consumption.

Endogenous factors can also be derived from the socio-cultural environment, as exemplified by changes in the daily routine and mode of life, cultural patterns of feeding, as well as from the technical environment, as demonstrated by new chemicals in food and medicine that are mutagenic. Essentially, none of these factors occurs in isolation; typically, they act in groups and interact with each other, and this can sometimes reinforce their effects. Alone among many environmental factors, interactions should receive special attention with reference to the etiology of diseases and depression, as only this knowledge can be the basis for appropriate preventive and therapeutic recommendations.

The history of mankind has witnessed several important forms of civilisation, of which only a few have survived. Presumably, cultural disturbance of organism–environment interactions (nowadays termed diseases of civilisation) already disrupted some earlier civilisations, or even accounted for the extinction of those cultures and their representatives. Thus, diseases of civilisation in the industrial age would be nothing unusual, and the history of mankind shows to what degree they can threaten a given civilisation. However, past civilisations were local in character, whereas the urban-industrial civilisation is a global event; hence the whole of mankind is endangered.

It seems that the main cause of contemporary diseases is distress (specific stresses were present in each civilisation) attributable to growing population density and excess of information, together with pollution of the atmosphere, waters, and soil with pesticides, residues of artificial fertilisers, and industrial and communal wastes. Diet, and contamination of food, may also play a very important role.

Acknowledgement

The last section of this chapter is co-written by Jerzy Rzepka.

Glossary

Acclimation   Temporary, reversible adjustment to climatic conditions only (acclimatory **Adjustment** also covers the impact of some other factors).

Acclimatization   Evolutionary (genetic) adaptation (e.g. skin or hair colour) to a given climate; in ontogenetic sense: 'preadaptation' to a particular climate (niche); [regulatory adjustment to climatic conditions].

Adaptability   Adaptive capacity; genetic plasticity; capacity of the organism [population, society] to respond to environmental pressures through the maintenances of biological homeostasis and cultural adjustments (behaviour) adequate to the environmental conditions; capacity for the maintenance of organism–environment equilibrium; possibility for cultural adaptation to a particular environment; capacity of a population to respond by genetic changes or plastic adjustments in biochemical, physiological, behavioural, and other mechanisms.

Adaptation   Any characteristic which enhances an organism's survival and reproduction; structural and functional characters of organisms, which enable them to cope with their environment; evolutionary process in which organisms grow to concordance with their environment. **Genetic adaptation** involves evolutionary changes in gene pool between generations (population's genetic constitution) adequate to environmental conditions, involving selection (negative = differentiated mortality and/or positive = differentiated fertility), which enhance the biological success of a population; biological success of the population. **Cultural adaptation** entails institutional (cultural) changes based on cognitive experiences related to environmental conditions and transmitted by teaching to next generations; formation of traits by a factor.

Adjustment   A temporary phenotypic change (physiological, structural, behavioural) to achieve organism–environment equilibrium (maintain homeostasis). **Regulatory adjustment** [reaction] = short-term reversible, mostly defensive physiological and/or behavioural changes. **Acclimatory adjustment** [habituation] = long-term reversible changes related to existing climate, nutrition, etc. (e.g. acclimatory changes in lung functions, blood morphology; functional hypertrophy of soft tissues); **Homeorhesis. Developmental adjustment** [plastic changes] = non-reversible changes in ontogenetic development related to niche, not transmitted in a biological way to later generations; ontogenetic **Homeorhesis. Cultural adjustment** = adaptive behaviour including products and their uses: machines, clothing, diet, settlement patterns, shelters, etc.; connected with any non-biological changes in humans instead of biological changes important for survival and development.

Capacitance   Exchange capacity; stress tolerance; limits to information exchange between an organism and the environment; tolerance for inputs and outputs.

Distress   Excess of stress, stress in negative sense; stimuli over **Capacitance**.

Evodeviation   Changes in conditions under which a species originated and developed into particular genetic form.

Genotype   A set of genes of an organism, genetic constitution of an organism (subject), opposed to **Phenotype**.

Heterosis   Effect in offspring of mating between parents of different genotypes; may result in a high hybrid vigour, high **Ecosensitivity**, etc. See Table 4.1.

Heterozygous   Of or pertaining to a cell (organism) with different alleles in a locus (loci) of homologous chromosomes.

Homeorhesis   Dynamic equilibrium of an organism in ontogenesis; ongoing (changing) **Homeostasis**; maintenance of biological (ecological) equilibrium in time; applicable to both organism and population.

Homogenousness   Status of a population or group with dominance of homozygous subjects.

Homeostasis   (Maintenance of) steady state of organism and/or population; intrinsic balance due to feedback responses; constant status of organism in fluctuating environmental conditions.

Homosis   Effect in offspring of mating between genetically similar parents; reciprocal effect to **Heterosis**.

Homozygous   Of or pertaining to a cell (organism) with identical alleles in the two corresponding loci of a pair of chromosomes.

Human biology   Branch of biology related to genetics, ecology, auxology, etc. of man; scientific study of genetic and ecological interactions in human phylogenetic (evolution) and ontogenetic development, and contemporary morphological, physiological and psychomotor variability of Homo sapiens.

Human ecology   Scientific study of man and his culture as a dynamic part of ecosystems; term most probably used first by Huntington (1916) in geography, next by Park (1921) in sociology, for special trends or doctrines in the above disciplines; first stage was monodisciplinary (as above, but also in anthropology, medicine-epidemiology, archaeology, economics, etc.), the next multidisciplinary stage was a mosaic of knowledge about man and his environment from these same disciplines; contemporary transdisciplinary stage means a synthetic knowledge of man and his culture as a part of ecosystems, based on classical disciplines. **Human ecology** has at least four main branches: philosophical and other theoretical aspects of interrelations between man and the environment; biological and social problems of human environment; ecological problems of human biology; and cultural adaptive behaviour, including environmental education.

Maladjustment   inadequate or poor **Adjustment**; organism–environment imbalance (lack of harmony); changes under conditions of **Evodeviations**, or negative side effect of intentional changes.

Modifier   (modificator, ecological factor)   Factor exceeding limits of habitual variation and tolerance of an organism in the previous period of development; an extreme factor which causes a change in an organism.

Paragenetic factor   Phenotypic variability is dependent on four groups of factors: genetic (endogenous), paragenetic, lifestyle and environmental (exogenous). Paragenetic factors are mixed endogenous–exogenous factors influencing embryo and foetus during intrauterine life: genetic resonance of foetus is not transmitted from maternal genes but is created by maternal intrauterine environment, which depends also on age of mother and number of previous pregnancies.

Pejus   Zone of physiological stress, rarely settled by a species; transitory condition between optimum and **Pessimum** for a species.

Pessimum   Area inadequate for a species, lowering physiological functions, the last zone of survival; man is not biologically adapted to this condition, but protected by cultural adaptation (man-made environment) and also adjusts biologically.

Phenotype   Observable and measurable trait or set of traits of an organism (subject); as opposed to **Genotype**.

Polygenic trait   Determined by interaction of more than one gene (several genes); mostly a continuously distributed trait.

References

Alexeeva, T.I., 1986, *Adaptational Processes in Human Populations*, Moscow University Publishing House, Moscow. In Russian.

Chrzastek-Spruch, H., Wolanski, N. and Wrebiakowski, H., 1984, Socio-economic and endogenous factors in growth of 11-year-old children from Lublin, *Collegium Antropologicum*, 8(1): 57–66.

Koziol, R., 1983, Hemoglobin concentration, hematocrite index and erythrocyte (RBC) number in some Polish populations. In: *XX Annual All-Poland Paediatric Conference*, 1: 67–69.

Koziol, R., Relationships between respiratory, cardiovascular and blood traits in various rural, industrial and urban populations, *Studies in Human Ecology*, 8: 95–117.

KoziolKolodziejjska, R., 1992, Respiratory-cardiovascular adaptation-adjustment in 7–49 years old inhabitants of selected regions of Poland, *Studies in Human Ecology*, 10: 221–232.

Leonetti, D.L., 1976, *Fertility in Transition: An Analysis of the Reproductive Experiences of an Urban Japanese–American Population*, PhD Dissertation, University of Washington.

Lindgren, G., 1976, Height, weight and menarche in Swedish urban school children in relation to socio-economic and regional factors, *Annals of Human Biology*, 3: 501–528.

Miklaszewski, W., 1912, Body development of Warsaw proletariat in the light of anthropological measurements, Warszawa (in Polish).

Morabito, F., Nicoletti, I., Marchi, M., Marella, M. and Bianchi, F., 1982, On the auxological interpretation of percentiles and of the area of normal growth, *Acta Medica Auxologica*, 14: 149–158.

Nicoletti, I. and Pelissero, G., 1979, Considerazione su uno standard di crescita staturale per la popolazione italiana, *Acta Medica Auxologica*, 11: 117–126.

Peters, J., Hashim, S., Marshall, W.A., 1982, Do we need local growth studies? *Abstracts of Conference of Society for Study in Human Biology*.

Pyzuk, M. and Wolanski N., 1972, Respiratory and cardiovascular systems of children under various environmental conditions. PWN, Warszawa (in Polish).

Siniarska, A., 1994, Rozwoj biologiczny dzieci i mlodziezy z kilk wybranych regionow Polski na tle warunkow zycia rodziny i pewnych cech biologicznych rodzicow, *Polskie Studia z Zakres Ekologii Czlowieka*, Supplement 1 to *Studies in Human Ecology*, pp. 89–194.

Siniarska, A., 1996, Family environment, parents' constitutional characteristics and biological development of children, in *The Family as an Enviroment for Human Development*, N. Wolanski and B. Bogin (Eds), Kamla-Ray Enterprises, Delhi, pp. 145–183.

Tanner, J.M., 1981, *A History of the Study of Human Growth*, Cambridge University Press, Cambridge.

Wang, R., Niu, T. and Shi, Y., 1996, Family transition and its human ecological implication in China, in: *The Family as an Environment for Human Development*, Wolanski, N. and Bogin, B. (eds), Kamla-Ray Enterprises, Delhi, pp. 101–111.

Wolanski, N., 1959, Essence of physical development of man and problems of its evaluation, *Kosmos*, 6(41): 601–616. In Polish.

Wolanski, N., 1961a, Normality, regularity and pathology from an anthropological point of view, *Czlowiek w Czasie i Przestrzeni*, 4(16): 171–175. In Polish.

Wolanski, N., 1961b, A new graphic method for the evaluation of the tempo and harmony of physical growth of children: The method of developmental channels and steps, *Human Biology*, 33(4): 283–292.

Wolanski, N., 1965, *Methods of Checking of Physical Development of Children and Youth*, Polish Medical Publishing House, Warsaw.

Wolanski, N., 1968, Biological reference systems called normal values and their practical significance in pediatrics, *Pediatria Polska*, 43: 775–783. In Polish.

Wolanski, N., 1971, About the theory of the limited direction of development, *Acta Medica Auxologica*, 3: 201–215.

Wolanski N., 1972, Human developmental stages connected with sensitivity to developmental factors, in: Wolanski, N. (ed.), *Factors of Human Development, Introduction to Human Ecology*, Polish Scientific Publishers, Warszawa, 483–530. In Polish.

Wolanski, N., 1974, Biological reference systems in the assessment of nutritional status, in: Roche, A.F., Falkner, F. (eds), *Nutrition and Malnutrition*, Plenum Press, New York, pp. 231–232.

Wolanski, N., 1975, *Methods of Checking and Standard Values of Growth and Development in Children and Youth*, Polish Medical Publishing House PZWL, Warsaw. In Polish.

Wolanski, N., 1975, Human ecology and contemporary environment of Man, in: Introductory Remarks, and Lectures, *14th Yugoslav Congress of Anthropologists*, Zagreb.

Wolanski, N., 1977, Genetic and ecological control of human growth, in: *Growth and Development: Physique, Symposia Biologia Hungarica*, 20:19–33, Akademiai Kiado, Budapest.

Wolanski, N., 1983, *Environmental Determinants and Biological Development of Man*, Ossolineum, Wroclaw. In Polish.

Wolanski, N., 1983, *Health – Environmental Control and Positive Indices*, Zdrowie Publiczne, 94(5): 241–258. In Polish.

Wolanski, N., 1990, Human population as bio-indicator of environmental conditions: Environmental factors in biological status of population of Poland, *Studies in Human Ecology*, 9: 295–321.

Wolanski, N., 1991, Human ecology and problems of demography, *Collegium Antropologicum*, 15(1): 27–43.

Wolanski, N., 1994, Monitoring program of biological status of human populations related to environmental changes, *Studies in Human Ecology*, 10: 113–139.

Wolanski, N., 1993, Selected problems of human ecology related to town planning, *Acta Oecologie Hominis*, 2: 1–194, Lund.

Wolanski, N., 1994, Srodowiskowe i cywilizacyjne zagrozenia wspolczesnego czlowieka, *Polskie Studia z Zakresu Ekologii Czlowieka*, Supplement 1 to *Studies in Human Ecology*, 195–224.

Wolanski, N., 1994, *Family as micro-environment of child development*, Final report for CONACYT project No. 1324–S9206, CINVESTAV, Merida.

Wolanski, N., 1995a, Household and family as environment for child growth: Cross cultural studies in Poland, Japan, South Korea and Mexico, in: Wright, S.D., Meeker, D.E. and Griffore, R. (eds), *Human ecology: Progress through integrative perspectives*, Society for Human Ecology, Bar Harbor, Maine, 140–152.

Wolanski, N., 1995b, Modernization as a form of cultural adaptation to the environment, in: Vatsyayan, K. (ed.), *Prakarti: the Integral Vision*, Vol. 5: Man in Nature, Printworld, New Delhi, pp. 245–258.

Wolanski, N., 1995c, Offspring stature and family factors in Japan, Korea and Mexico: multiple regression analysis, *Ninth Congress of European Anthropological Association*, Copenhagen, in press.

Wolanski, N., 1996, Household and settlement as environment of human development in contemporary civilisations (introductory remarks), in: *The Family as an Environment for Human Development*, Wolanski, N. and Bogin, B. (eds), Kamla-Ray Enterprises, Delhi, pp. 15–41.

Wolanski, N. and Antoszewska, A., 1990, The diagnostic value of somatic traits related to their rate of growth in the pre- and postnatal period, *Endokrynologia Polska*, 41(3): 159–171.

Wolanski, N., Chrzastek-Spruch, H., Kozlowska, A., Teter, A. and Siniarska, A., 1988. The role of culture, living conditions and genes in the growth of 11-year-old children from Lublin, *Antropologia Contemporanea*, 11(3–4): 167–175.

Wolanski, N., Chung, S., Tsushima, S., Tomonari, K. and Czarzasta, T., 1994a, Family types and offspring growth in various countries: III, Regression of offspring's stature in relation to parents' and family's factors in Japan and Korea, *Studies in Human Ecology*, 11: 23–29.

Wolanski, N., Czarzasta, T., Chung, S., Tomonari, K., Tsushima, S. and Seiwa, H., 1994b, Family types and offspring growth in various countries: II, Stature of offspring in various family types in Japan and Korea, *Studies in Human Ecology*, 11: 13–22.

Wolanski, N. and Januszko, L., 1990, Variation of women's fertility and survival of their offspring, *Studies in Human Ecology*, 9: 213–224.

Wolanski, N. and Siniarska, A., 1984a, Species module and assortative mating in Man, *Journal of Human Evolution*, 13: 274–253.

Wolanski, N. and Siniarska, A., 1984b, Unreliability of some biological criteria for evaluation of the suitability of urban ecosystems for Man, Garcia de Orta, *Seria Antropobiologia*, Lisbon, 3(1–2): 15–18.

Wolanski, N. and Takai, S., 1976, Age changes in asymmetry of distribution of some somatic biochemical and psychomotor traits, *Acta Facultatis Medicinae Universitatis Brunensis*, 57: 377–388.

Wolanski, N. and Zaremba H., 1996, Regression of offspring morphology to socio-economic and genetic family factors in rural and urban populations of Poland, *Collegium Antropologicum*, 20(1): 37–52.

Wrebiakowski, H., Chrzastek-Spruch, H., Wolanski, N., 1982, Profession, education and income of parents, and growth of 11-year-old children from Lublin, in: *XX Annual All-Poland Paediatric Conference*, 1: 57–60.

5 Health through sustainable development

A potential planning contribution

Jerzy Kozlowski and Greg Hill

Defining sustainable development

Sustainable development has recently become one of those magic terms used by almost everybody though not necessarily in a proper way or in the right context. Some people believe the concept to be a sort of 'philosopher's stone' which will solve all our problems. Others tend to be suspicious as to whether it really does mean anything.

To address these doubts some basic concepts of a new 'development philosophy' must first be discussed as they have opened up new horizons for the future of our planet. It has been intimated, for instance, that development need not necessarily be equated with growth. The two terms are not synonymous although often, and wrongly, they are used interchangeably. Development means the realisation of specific social and economic goals which may call for a stabilisation, increase, reduction, change of quality or even removal of existing uses, buildings or other elements, while simultaneously (but not inevitably) calling for the creation of new uses, buildings or elements. It must be noted that in each case development should lead to progress, expressed primarily by welfare improvements in the communities involved, and that it will occur through specific changes (Kozlowski, 1986; Zarsky, 1990). Such improvements do not necessarily result from growth itself. Therefore, a no-growth situation does not mean that there is no development. So long as there is the need for progress there will be the need for development and, in turn, the need to decide where and how this development is to take place so that the cost of it can be minimised and the benefits maximised. Here, cost needs to take on a broad meaning, that is, not only economic but also ecological and social. As this cost is, among others, strongly influenced by location, scale, kind and timing of development it is imperative that their consequences on the environment and on the human communities involved will always be established early in the decision making processes.

The obvious urgency for a new statement of a 'true' (sustainable) national income is addressed by the recently launched, *Strategy for Sustainable Living* which defines it as:

> Net National Product (Gross National Product minus depreciation of human-made capital) plus increases in natural assets minus depreciation of natural

assets minus defensive expenditures against environmental damage minus the costs of unmitigated environmental damage.

(IUCN *et al.*, 1991: 74)

All this confirms the view that, opposite to what some believe, the development problems cannot be solved solely by the presently operating free market mechanism which is extremely efficient in the optimum allocation of scarce resources but which fails to address the cost of depreciation and use of such basic natural resources as air, water or soil. It will also not provide the answer to an elementary but essential question: How much enough is enough?

The reason is simple. Conventional free market economists do not recognise that there are 'limits' to economic growth and believe it can go on exponentially forever. At the same time most of the main advocates of sustainability (Daly and Cobb, 1989; Pearce *et al.*, 1989; Barrow, 1995) agree that there are *final limits* to what the natural environment can take, that these limits cannot be continually violated without a threat to our survival, and that science or technology can never provide effective means of permitting the extension of these limits indefinitely.

Having exposed and addressed these fundamental issues, the late 1980s and early 1990s have, among other things, shone new light on the concept of *sustainable development* (IUCN *et al.*, 1980) which indicates how both economic and ecological sustainability may be integrated to produce mutual benefits. Its primary goal is to achieve a reasonable and equitably distributed level of economic well-being that can be perpetuated through 'development that meets the needs of the present without compromising the ability of future generations to meet their own needs' (WCED, 1987: 43). This is, perhaps, the most frequently quoted (among nearly a hundred) definition of sustainable development. Pearce and colleagues have a similar view, when seeing the sustainability criterion as the one requiring 'that we leave the next generation a stock of "quality of life" assets no less than those we have inherited' (Pearce *et al.*, 1989: 34). Sustainable development is also defined as a pattern of social and structural economic transformations which optimises the economic and societal benefits available in the present, without jeopardising the likely potential for similar benefits in the future (Goodland and Ledoc, 1987), and the sustainable society as the one that lives within the self-perpetuating limits of the environment. It is not a 'no-growth' society but, rather, a society that recognises the limits of growth and looks for alternative ways of growing (Coomer, 1979). To the 'Strategy for Sustainable Living' it means, in turn 'improving the quality of human life while living within the carrying capacity of supporting ecosystems' (IUCN *et al.*, 1991: 10).

Sustainable development and sustainable health

Virtually all the above definitions stress life quality, human well-being, or welfare, as the ultimate aim. This concept is, therefore, clearly related to human health, both in the physical sense in so far as it relates to adequate food and shelter or protection from disease and contamination, and as well in a spiritual sense (where these broad notions need to be defined in terms of the adequacy of the environment to provide for emotional well-being). Yet, it is quite amazing that despite the conscious linking of sustainable development with human welfare, so few authors see its logical link with 'human health'. One is Tolba (1987). He lists health control as one of the five main issues that sustainable development must embrace. Pearce and colleagues also understand that life quality depends greatly on the health of the population; and that better environmental quality, achieved through sustainable development, would frequently improve economic growth by, among other things, 'improving the health of the workforce' (Pearce *et al.*, 1989: 21). More recently, in their 'manifesto for living' Pickering and Owen see, as one of their ten goals to attain more sensitive living with the natural environment, an improvement of basic medical care. They argue that:

> longevity and good health correlate extremely closely with wealth. The richer nations enjoy the standards of health provision which is far above that available to all but a few in the poorer nations. The imbalance of the most fundamental human provision, after food, should be rectified through greater international direct aid to countries where health care is limited.
>
> (Pickering and Owen, 1994: 317)

Looking at the enormous publicity that sustainable development has received in recent years, and particularly after The Brundtland Report, *Our Common Future* (WCED, 1987), it is difficult to say why human health has been so conspicuously absent in the main stream of discussion on and struggle for, sustainable development and somewhat surprising how little attention has been give to the concept of '*sustainable public health*'. This is despite the fact that threats to human health from changes to the global environment, such as ozone depletion, habitat destruction, species extinction, global warming, and the poisoning of air, water and soil are known and well documented (Chivian *et al.*, 1993).

They directly concern our own welfare and, perhaps, even survival. The fact that any success in dealing with environmental dangers to human beings strongly depends on success in achieving sustainable development has not received satisfactory attention.

Are we too frightened of the scenarios to concentrate on the very real implications of environmental degradation on human health and survival? Or is it, perhaps, less threatening to concentrate on elements of the natural environment and their demise? These questions, surely, do reflect on *the disciplines of* ecology and planning. It may, therefore, be worth exposing the interrelationships between sustainable development and sustainable health a bit further.

'*Live and die healthy*' may be a simplistic definition of sustainable public health. As simplistic as the answer given by a man celebrating his 100th birthday and asked what advice he could give about living a long life. He replied: 'keep breathing'. But if one adds 'keep eating and keep sheltering' the main cornerstones of sustainable health would be listed. Certainly with the *caveat* that what we eat and breathe is not poisonous and polluted.

During the national, participative workshop '2020: a sustainable healthy future – towards an ecology of health' (held in Melbourne, Australia, in April 1989) its chairperson Dr Ilona Kickbusch defined public health itself as:

> the science and art of promoting health. It does so based on the understanding that health is a process engaging social, mental, spiritual and physical well-being. It bases its actions on the knowledge that health is a fundamental resource to the individual, the community and to society as a whole and must be supported through sound investments into conditions of living that create, maintain and protect health . . . Public health is ecological in perspective . . . [and] . . . aims to improve the health of communities through an organised effort based on:
>
> - 'advocacy' for healthy public policies and supportive environments
> - 'enabling' communities and individuals to achieve their full health potential. . . .
>
> Public health infrastructures need to reflect that it is an interdisciplinary pursuit with a commitment to equity, public participation, sustainable development.
>
> (La Trobe University 1989: 2)

The workshop recommended, among the criteria proposed for healthy public policies, that:

- sustainability, understood as the capacity of social and economic developments to meet the needs of the present generation without compromising the ability of future generations to fulfil their own needs, must become the central criterion for social and environmental impact statements; and that
- the criterion for evaluating the impact of policies, in all sectors, on the global environment, should be in terms of effective utilisation of resources without disruption to ecological balance.

In trying to address these two broad issues, one of the main questions discussed at the workshop was: what strategies do we have, to ensure we live in an ecological balance which does not threaten our quality of life?

The question conforms with most definitions of sustainable development which stop short of making a direct link between life 'quality' or human 'welfare' and human health. To seek solutions to any problem by asking questions is always most appropriate. The ability to formulate the right questions can be more important than finding right answers. Once the right question is asked, sooner or later, someone will come up with the right answer. 'Right asking' includes the ability to listen to questions offered by others. Consequently, new formulations can be created and necessary refinements made. Above all, however, questions should be specific rather than general. Certainly not as general as those centred on 'quality' or 'welfare'. Inquiry at this level can include 'everything' and hence lack any specificity at all. Frequently, such motherhood statements are later skilfully abused and spread by politicians and decision makers.

Hence it is essential to examine whether and how the notion of health would entwine with the specific meanings of quality of life, human welfare and well-being.

Clearly these meanings must reflect and be directly related to the degree of satisfaction of 'human needs' and to their general classification.

Satisfaction of these needs was always behind any development in the history of humankind and in particular, behind the development of human settlements. Providing shelter and security, for instance, was essential from the dawn of civilisation. Over time, improvements in sanitary conditions, widening economic opportunities, accessibility, or choice rose to prominence in the continually expanding list of human needs.

Overarching the coherent typology of needs, it is proposed to follow Gorzelak (1986) and to consider the 'needs' from two main view points, namely:

* external viewpoint, which expresses an objective definition of needs as assessed by an outside observer, and thereby indicate '*conditions of life*'; and
* internal view point, meaning the subjective identification of need based on self-assessment, by an individual or by a particular concerned community, and which would indicate, in turn, what is being perceived as the '*quality of life*'.

Human needs may be further categorised into two physical and intellectual domains:

* Physical or material needs: that is those needs whose satisfaction involves direct input of material goods; such as the need of shelter and food and the provision of sanitary conditions. Our 'existence' is determined by their fulfilment.
* Intellectual or spiritual needs: that is those needs whose satisfaction does not depend on a direct input of material goods; such as the need for justice, democracy and individual and community participation. Hence 'being' is determined by their fulfilment.

A matrix of this basic typology is provided in Table 5.1.

Such an approach implies that, even at the most general level of discussion about health, one must address not only the 'quality' of life itself but the 'conditions' of life as well. Problems occur when satisfaction of a particular need is hindered by specific difficulties. Referring to a definition by Chadwick (1971), a 'problem' is seen as a question or matter proposed for solution which equals needs plus difficulties in their attainment. A meticulous and precise definition of these human needs that regard sustainable health must, therefore, precede the definition of problems to be solved, and must become an inherent part of the generation of '*development aims*' in the decision making process. Only then would sustainable health interests be properly safeguarded.

Table 5.1 General classification of human needs

	Physical needs (material input)	Intellectual needs (no material input)	
Objective	Conditions of existence	Conditions of being	Conditions of life
Subjective	Quality of existence	Quality of being	Quality of life

In general, any basic aim of development directed at the satisfaction of human needs should be derived from the aims of human settlements which, according to a UNEP definition, are:

> to meet needs and aspirations of communities living in these settlements by providing conditions both for proper functioning of these settlements and for the biological, social, economic and intellectual development.

(UNEP, 1977)

Simonis (1990) reminds that, according to the definition adopted by the World Employment Conference in 1976 (confirmed by the UN General Assembly) basic needs, listed under Article 2, embrace among others: the provision for regular, minimum requirements of food; housing; clothing; and such vital services as safe drinking water and health care. This follows the earlier, Drewnowski indicator-system, where health is one of the nine components which should not be absent from any measurement of the level of living (Drewnowski, 1974).

Thus, the fundamental aims of development necessitate the need to:

- secure the 'survival' of a given settlement, or rather the community living in it, by defining how its environment is to be protected (Conditions for Existence); and
- seek out an urban form (pattern, strategy) such that an optimum basis be created for the 'functioning' of a given settlement and for its biological, social, economic and intellectual 'development' (Conditions of Life and Quality of Life).

Sustainable health, in this perspective, would rely above all on satisfying conditions for existence and, most importantly, would be intrinsically linked with the fundamental question of survival. Whether it would be achieved, greatly depends, in turn, on achieving both 'economic' and 'ecological' sustainability of development. As a consequence, it seems imperative to address them jointly whenever possible. This view is shared, by Zarsky (1990), among others, who when addressing one of the two main challenges faced by the Australian National Strategy for Ecologically Sustainable Development, specifies the need to reverse the deterioration of the environment and, significantly, states that if this fails 'our health and livelihood, as well as that of future generations, hang in the balance' (Zarsky, 1990: 1).

Particularly solid evidence documenting intrinsic relations between human health and the natural environment is provided in a Report by Physicians for Social

Responsibility (Chivian *et al.*, 1993). Two of the three main themes underlying the report, according to Cortese in his introductory chapter, are:

• The physical environment, our habitat, is the most important determinant of human health.
• Protection of the environment and preservation of ecosystems are, in public health terms, the most fundamental steps in preventing human illness.

(Cortese, 1993: 1)

Given these close connections between environment and health, Cortese expresses doubt about how health can be promoted within a fundamental belief system that prevents us from recognising the dimensions of our environmental crisis. He argues for stopping further environmental degradation as a first step towards a solution. It can be achieved by implementing major shifts in policies that seek to control and prevent such deterioration in the environment. Secondly, he advocates changing relations between developed and developing countries by the implementation of 'sustainable development'. That must be firmly based on a profound understanding of the interdependence of all nations when it comes to the global environment.

Most of the evidence presented in the report is disturbing – particularly the facts revealing the number of people whose deaths are directly attributable to gross pollution of air and water. Many of these are children (Christiani, 1993; Leowski, 1986).

As defined by many reports and studies (Thurston, 1989; Ostro, 1984; Holland and Reid, 1965; Mazumdar *et al.*, 1982; Niki *et al.*, 1972) 'atmospheric pollution has now reached a level that threatens not only the health of entire populations but also their survival'(Christiani, 1993: 26). These facts are often matched by those in other parts of the report, particularly those dealing with dirty water, food contamination, radiation, the ozone layer, climate change and population growth.

The summary chapter by Chivian (1993: 193–224) is very telling as it highlights the fact that human activities are causing the extinction of species at rates which are thousands of times those that would have occurred naturally. Some 27,000 species are lost in tropical rainforests alone each year (a rate of three per hour) and even those figures may be underestimates (Wilson, 1992).

Unfortunately, it is not realised how such massive extinctions may limit prospects for finding treatments to many terminal diseases threatening human survival and causing enormous human suffering such as cancer, AIDS, or arteriosclerotic heart diseases. Clearly, preventing further loss of biological diversity by recognising the importance of achieving sustainable development, can be seen as nothing less than a medical expediency for the human race.

Chivian argues convincingly that the destruction of habitats and the loss of species may upset the delicate equilibria among ecosystems on which all life depends. Particularly, this could be the case with regard to food supplies which may be significantly threatened by factors such as deforestation, and resulting rainfall reduction, or topsoil erosion. Both of these are capable of compromising crop production. Overfishing and the destruction of coastal wetlands, nurseries for many species, have serious implications for the human race. So do factors related to the spread of infectious disease, which may increase substantially following damage to the ecological balance between hosts and parasites and between predators and prey, disturbed by habitat destruction or by loss in biodiversity.

The crux of the reports findings state:

> Human activity is causing the extinction of animal, plant, and micro biological organisms at rates that may well eliminate one-fourth of all species on Earth within the next 50 years. The incalculable human health consequences from this destruction include . . . the upsetting of the balance among ecosystems on which all life, including human life, depends. Major efforts to protect natural habitats and to preserve biodiversity are required to prevent these medical catastrophes from occurring.
>
> (Chivian, 1993: 218–219)

Similar concerns can be also found in the recent, major IUCN publication where Rodgers and Saunier (1994) in their 'big picture' state that basic health care is lacking for some 1.5 billion people, while over 1.5 billion do not have safe water and over 2 billion lack satisfactory sanitation. Is it not astonishing that the magnitude, importance and extreme urgency of problems faced in the pursuit of sustainable public health do not receive satisfactory global promotion and seem to be indeed marginalised?

This can be illustrated by the Melbourne workshop (1989) which received little publicity or by the acclaimed Earth Summit in Rio de Janeiro 1992, where the prospect of global warming was put far ahead of the continuing massive loss of human life from water and air pollution in Third World countries. The focus of mainstream environmental movements worldwide seems misplaced and thus becomes one of the major problems in itself to be addressed.

Should not sustainable public health, therefore, become one of the priority goals of sustainable development? In selecting a pathway towards it, the community tends to be influenced by its own aspirations which often change dramatically over time. In the 1950s, for example, many people dreamed of owning a house in the suburbs, and freeway links were provided accordingly. Today many people may prefer, in turn, to live in townhouses close to central city areas. There are, however, community goals that would never change. The recent Handbook on 'Strategies for National Sustainable Development' (Carew-Reid *et al.*, 1994) put forward six such 'universal' goals: a long and healthy life is the first, followed by education, access to resources ensuring decent living standards, political freedom, human rights and freedom from violence. Whether long and healthy life is achieved depends strongly on the success of sustainable development. As a consequence, this goal should become one of the priority goals as well and included in its definition. Sustainable development could then be seen as:

development leading towards improving human health and welfare for the present generation, while being contained within the carrying capacity of life supporting ecosystems to ensure that the ability of future generations to achieve the same goal is not compromised.

The evidence that health is a fundamental resource to the individual, the community and to human society is overwhelming; as is the fact that social, spiritual and physical well-being depends on the ecological status of the environment in which society functions. Sustainable development, therefore, not only has relevance for the status of natural ecosystems, but may also provide one of the cornerstones for the achievement of sustainable public health.

If there is to be a sustainable, healthy future for people on 'spaceship Earth', it is imperative that governments and communities at all levels ensure that their policies and strategies take into account health impacts and address the question of how to contribute towards achieving an ecological balance which does not threaten the conditions and quality of life. The same question should be directed to all professional disciplines including the professional discipline of planning.

Planning for sustainable health: the potential

The importance of the role of planning in achieving sustainable development is underestimated and the concept of sustainable development has yet to influence the planning profession. The concept of sustainable development and its implications calls for a fundamentally different approach to development from that provided by the standard economic model which has dominated planning practice for so long. The challenge facing managers and planners, then, is to utilise the principles of ecologically sustainable development as an integral part of the planning process.

Important questions arise.

Could this 'different approach' suffice for planning to make a marked contribution towards achieving sustainable public health in the light of the paramount problems briefly outlined above?

How could sustainable development influence the planning profession and, in particular, its day-to-day approach to the preparation of statutory plans of various kinds, planning studies, strategies or guidelines?

The 'ecological' role and responsibility of professional planning is clear. Planners should ensure that decision makers (primarily politicians and developers) are fully informed about all environmental consequences of development policies or projects. In this way planning can contribute to the prevention of further degradation of the natural environment in the field which influences it most, that is, in decisions on the location, scale, kind and timing of development. Planning must be anticipatory and proactive.

Is the planning profession fulfilling this major responsibility?

The everyday approach of the planning profession to environmental problems is still conservative. That is, '*ex post*' instead of '*ex-ante*', as it deals primarily with attempting to cure the symptoms rather than preventing the causes. Economic utility and political expediency most commonly determine the location, scale, kind, and timing of development; ecological and aesthetic utility or sensitivity of natural resources are usually regarded as externalities and hence conveniently consigned to oblivion. It is forgotten (or not realised) that natural resources provide crucial *'environmental services'* such as maintenance of water flow patterns, protection of soil, bio-degradation of pollutants, recycling of wastes, regulation of climate, support of fisheries and other economically important living resources. Such interrelationships between people and nature are frequently misunderstood or ignored by the majority of developers, governments and even professional planners. Sooner or later such attitudes will lead directly to both ecological and economic disasters.

A central finding of a review of strategic planning in Queensland, Australia, confirms that such an 'ecological' way of thinking has yet to penetrate professional planning (Kuiken, 1990; Kozlowski, 1990). It is hard to identify a specific planning approach, *commonly used in practice*, which properly deals with, or even reflects, the requirements and objectives of sustainable development. Judging by available published materials and discussions taking place at various international planning conferences, it can be inferred that most statutory planning documents, in most countries, reveal similar deficiencies. This weakness seems to be global in character.

The conclusion is clear: it is a matter of urgency to establish principles and working tools capable of effectively addressing environmental threats within planning and development decision-making processes.

It can be argued that Environmental Impact Assessments (EIAs) in their diverse forms have provided the above by injecting a badly needed 'preventative spirit' into the planning process. But, is this really the case? Consider these two central shortcomings:

- First, even in the few countries where EIA can be enforced, the lists of development projects for which they are legally required are particularly revealing, not so much for what is included but for what is not included. Developments which are potentially threatening to the natural environment and its resources such as agricultural types of development are usually not on such lists.
- Second and even more importantly, in the EIA procedure, measures mitigating adverse impacts on nature and natural resources are, at best, contemplated only in relation to the already allocated land uses. Furthermore, EIAs, developed and applied with only a modest involvement of the planning profession, have hardly ever become an integral part of the planning process. They represent, rather, a powerful but also independent and self-contained approach. If used at all in this process they are, as a rule, considered in its later stages, consistent with the nature of the EIA as 'essentially a project analysis technique' (Fowler, 1982: 285).

This means that the major planning decisions on the location, scale, kind and timing of development take place *before* any EIA involvement. Hence it is inevitable that development proposals are *generated* without ecological

scrutiny or perspective. Clearly, as far as the interest of the natural environment as a whole is concerned, the EIAs are not applied to planning and
decision making *early* enough. Some authors, in fact, find it astounding that
the implications of project location are not addressed earlier (e.g. Wathern,
1988).

The success and wide application of EIAs has overshadowed a fundamental short-
coming in day-to-day planning – the lack of environmental considerations early in
the planning process. 'Assessments' of all kinds, by their very nature, can only
appear in the later stages of the planning process. Development concepts, proposals, or projects must first be formulated in order to be assessed. Assessments by
themselves cannot influence the ways in which these concepts, proposals or projects are *generated*.

A similar view was strongly put in the previously mentioned publication by
IUCN edited by Munasinghe and McNeely (1994), where it is emphasised that
the only relevant entry point into the project cycle by an 'integrated planning'
approach to be of use

> is at a time . . . previous to the identification of projects. Rather than to estab-
> lish sectoral plans, full-blown projects, and environmental impact assessments,
> it first helps to define a development strategy, and then identify projects –
> *before* (emphasis JK and GH) investments of time, energy, funds, and politi-
> cal supports are made . . . [Such a] . . . multi-sectoral overview undertaken
> *early* (emphasis JK and GH) in the development planning process is the
> trademark of integrated regional development planning.
>
> (Rodgers and Saunier, 1994: 76)

Or, rather, 'should be' as this still seems to be one of the essential, and commonly
overlooked, 'environmental gaps' in the present planning approach. To bridge
such a gap could be an important contribution by professional planners towards
the attainment of sustainable development which is inextricably linked to sustainable public health.

An 'ecological' overhaul of the current planning approach, which aims to integrate ecological content into that planning approach right from the outset, while
retaining EIAs for the final stages, seems to be the most 'open' and promising
approach from a planner's perspective. Reform of the approach need not be dramatic as it can evolve through time, hopefully with sound co-operation between
planners, EIA and other environmental science experts.

By incorporating ecological considerations early in the planning process an
additional benefit of reduced demand for EIAs could be obtained. At present EIAs
are usually not only time consuming and costly, but 'conflict prone', especially
when commissioned by developers with vested interests. Planning, then, could
become an effective tool in achieving sustainable development.

What is suggested is surely common sense. It is, however, far from being universally accepted, let alone applied in everyday professional practice. Yet,

integration of environmental concern early in the planning and decision-making processes, before too much money is spent or positions are hardened, may help significantly to avoid future negative impacts and potential conflicts. Hardly anybody seems to realise such an obvious truism. It is particularly staggering when from the highest world decision-making level, the World Bank, comes a recommendation, regarding the Bank's Post-Rio Strategy for Sustainable Development, that

> the *first* (emphasis JK and GH) component of this strategy is the development of a comprehensive environmental assessment procedure which aims to ensure that development options under consideration are environmentally sound and sustainable.
>
> (El-Ashry, 1993: 22)

There seems to be no recognition, or acceptance, that the first component of the strategy should be the 'environmentally sound' process of *formulating* those options.

The potential role of planning has been, at last, realised by major proponents and advocates of sustainable development. Professional planning has been urged, for instance, to accept that development should follow a direct recommendation of the 'Strategy for Sustainable Living', which indicates the need

> to adopt and implement an ecological approach to human settlements planning to ensure explicit embodiment of environmental concerns in the planning process and thus promote sustainability.
>
> (IUCN *et al.*, 1991: 106)

A Handbook on 'Strategies for National Sustainable Development' (Carew-Reid *et al.*, 1994) has recognised planning as an important part of these strategies and it is also worth noting that planning, conspicuously missing in earlier drafts, has also found its way, as a separate chapter, into the recently published Australian *National Strategy for Ecologically Sustainable Development* (Commonwealth of Australia, 1992).

Concluding this brief overview of the potential planning contribution towards achieving ecologically sustainable development and, thereby, sustainable public health it should be remembered that development is determined by three major factors, namely:

- *social goals*, reflecting physical and intellectual needs of a given community;
- *geographic environment*, creating constrains and opportunities; and
- *development circumstances*, including the state of the economy, technology, social organisation, cultural tradition, political system, etc.

The role of planning in defining social goals can only be advisory as such goals should be defined by the community and development circumstances which are beyond a direct planning control. Planning, as a consequence, should concentrate on indicating how development can best proceed in a given geographical environment, to attain the goals within a set of existing circumstances. Therefore, planners designing or trying to determine the consequences of various development policies, proposals or projects, should concentrate upon implications derived from the geographical environment with particular attention being given to the two main issues of:

- *location, scale, kind and timing* of development, by defining the environmentally and economically sound 'solution space' for this development, derived from examining interrelations between development activities and available resources; and
- *form (patterns)* of development, by identifying a range of economically and ecologically reliable development options within this space.

This chapter focuses on the first issue, which reflects the previously identified environmental gap in present planning approaches and, as a consequence, on the need to introduce environmental considerations into the planning process right from its outset, at the stage where development policies, proposals or projects are generated. Thus, the emphasis in planning should move away from curing the symptoms to preventing the causes and to responding, as Barrow (1995) reminds us, to the generally more and more accepted requirement for an 'anticipate and avoid' approach to development, rather than a 'react-and-mend' approach (Barbier, 1987; *Futures*, 1988; Munn, 1988).

Planning for sustainable health: the tools

The discussion on sustainable development strongly implies that the recognition of final development limits, derived from the requirements of the natural environment, is the main step to be urgently taken. The same is true in planning because there is no use in dealing with the problem in a more detailed and sophisticated way before these limits are identified in the early stage of any planning exercise. Development must be containable within the carrying capacity of supporting ecosystems. Planning could make an important contribution if it is able to identify environmental limits that determine this capacity. And planning could make a further important contribution if it was able to ensure the rational and ecologically sound decisions concerning location, scale, kind and timing of such development. This, primarily, depends on the rational use of environmental resources.

Four principles for such use can be distinguished (Kozlowski 1986, 1993):

- The First Principle implies that respective activities should be developed primarily where, as a result of natural evolution and human development, there are the resources required by these activities. At the same time, the activities should be developed only where their side effects do not impinge on a sensitive facet of the environment. Thus, the advantages of a particular location should always be balanced against the total costs, including social and ecological costs, which may result from a given type of use.

- The Second Principle accepts that the rational use and exploitation of environmental resources require that economic activities be developed up to the levels indicated by quantitative constraints. These are determined both by the resource potential (size, yield, magnitude) needed for these activities and by the 'boundary degree', (degree beyond which the resistance may disappear), of resistance of those and other resources to negative side effects.
- The Third Principle concerns the quality of output of a given activity. The possible negative influence of this quality depends partly on the quantity and quality of required resources and partly on the technology of their exploitation and transformation.
- The Fourth Principle derives from considering the way in which the rate or duration of development is affected by the rate or rhythm of natural processes, leading to changes in quantity, quality and resistance of resources. The same can happen when the development does not provide for the necessary time intervals during which natural ecosystems can regenerate or re-establish themselves.

The 'Strategy for Sustainable Living' states, rightly, that 'the maximum impact that the planet or any particular ecosystem can sustain is its "carrying capacity"' (IUCN *et al.*, 1991: 43). Its definitions vary and can be imprecise. According to Barrow it may:

> include the maximum number of individuals that can be supported in a given environment; . . . the amount of biological matter a system can yield, for consumption by organisms, over a given period of time without impairing its ability to continue producing; the maximum population of a given species that can be supported indefinitely in a particular region by a system, allowing for seasonal and random changes without degradation of the natural resource base. For human society, there can be a maximum or an optimum carrying capacity.
> (Barrow, 1995: 58)

The capacity is ultimately limited by environmental constraints to accommodate development impacts and requirements. Any development proposals should, therefore, be considered within these constraints *prior* to decisions being made. The Four Principles indicate that there are four corresponding *dimensions* by which environmental constraints can be defined. These are, according to Kozlowski (1986):

- territorial;
- quantitative;
- qualitative; and
- temporal.

The definition of environmental constraints is of key importance at any planning level. They should play a critical role in establishing an ecologically sound 'solution space' within which development proposals would have to be generated and contained. This space, being the planning contribution towards defining 'carrying capacity', should primarily be determined by the constraints (derived from the four environmental dimensions) which can be considered as having final (boundary, ultimate) characteristics. These constraints should indicate where, at what scale, and which kind of development should take place, over what time period or at what rate, so that a rational use of natural resources can be secured. It must be kept in mind, however, that even those constraints and, therefore, carrying capacity can be stretched by technological measures and management practices, but 'usually at the cost of reducing biological diversity or ecological services. In any case, it is not infinitely expandable' (IUCN *et al.*, 1991: 43).

Definition of the constraints should take place early in the planning process, that is, at the stage when location, scale, and kind of development are normally decided.

This 'ecological' perspective evaluation of planning may be taken further in helping to address another generic problem faced by the planning profession which must continually contend with the dual, and ostensibly contradictory, responsibilities of *restricting* development and *promoting* development. This is because planning must both safeguard the conservation of nature and, at the same time, guide and stimulate socio-economic development. This contradiction can be dealt with by sub-dividing the planning process into two related strands – a 'restrictive' strand and a 'promotional' strand (Kozlowski, 1993).

The *restrictive strand* is the one in which priority is given to the conservation of nature and its resources. Within this strand, goals and objectives, which condition development, should be formulated. They would be, primarily, of an ecological character and would provide *a basis* for defining an environmentally sound 'solution space' for *formulating* development proposals. Undoubtedly, the key factors determining such a 'solution space', and the degree of suitability of specific areas for different kinds of development, should be environmental constraints. Such constraints should be supplemented by factors reflecting major values of the human-made environment, such as cultural or historic values. Therefore, 'solution space' is a physical and a cultural space which is regarded as appropriate for both economic and social development without, or with minimal, costs either to the environment or to the existing values of the society.

The *promotional strand*, in turn, emphasises a whole range of development options or scenarios to be constructed and fostered within the 'solution space' *determined* in the restrictive strand. These options should aim at fulfilling socio-economic goals and objectives, set at the beginning of this strand, and should be the subject of various environmental, social and economic assessments.

The essence of this approach is the recognition that a proper and reliable definition of the environmentally sound 'solution space' is perhaps the major responsibility of planning. The sequential integration of the restrictive and then the promotional strand is presented here as a crucial move towards planning for sustainable development. The defining of the 'solution space' must be based on knowledge derived from all major disciplines, having an affinity with development and its management; that is, not only architecture, engineering and economics, but also ecology, geography, biology, geology, botany and zoology.

Do we have a specific tool to use in the determination of 'solution space'? The technique known as the Ultimate Environmental Threshold (UET) method is suggested as being one such tool.

The Ultimate Environmental Threshold method

By its very nature, development causes changes in the geographic environment. These changes invite a two-fold conceptualisation:

* changes which result from the adaptation of the environment to the requirements of a given development, and which provide its *basis*; and
* changes which result from side effects, and which represent development *consequences*.

Such changes often encounter some physical limitations imposed by the geo-graphic environment. These limitations cause *discontinuity* in development processes expressed by slowing down or even stopping those processes unless the limitations are overcome. The overcoming of limitations would involve variable *additional costs* of development. Their magnitude depends on the location and the quantitative and qualitative levels of development and the costs which are adjusted accordingly to fit the basic features of the geographic environment. These costs, which may be high, would comprise not only investment costs but also social and ecological ones. The concept of discontinuity and of additional costs is charac-teristic of threshold-type phenomena and the limitations can therefore be considered as development thresholds (Malisz, 1969; Kozlowski and Hughes, 1972; Kozlowski, 1986). They occur when further development in a given area and within a given period faces either a decrease or absence of resources required by this development; or when such negative 'side effects' occur, causing damage both to the ecological balance of the environment and to the ecological functions of natural resources.

The following definition explains the concept further:

> a threshold to further development is encountered if it cannot extend to a new area, produce additional output, achieve higher quality, or accelerate produc-tion, without involving an increase of investment, social or ecological cost. The number of units of output of this development at which such a situation occurs indicates the threshold on a development curve (hypothetical or actual).
>
> (Kozlowski, 1986: 121)

The four major environmental constraints, previously mentioned, indicate the fol-lowing four corresponding types of development thresholds:

- *territorial*, indicating *location*, that is the area over which a given development can take place;
- *quantitative*, indicating *scale*, that is the level up to which the development can be carried out;
- *qualitative*, indicating *type* of development, that is the kind of output that can be accepted; and
- *temporal*, indicating *timing* of development, that is its admissible rate, or the permitted time periods in which it can take place.

The thresholds can be imposed directly by the potential of resources needed for a development, or indirectly by resources which can be seriously affected or damaged by its side effects. These often accrue and result in synergistic problems.

Since human settlements are totally dependent on the surrounding natural environment for the provision of resources and energy and for the disposal of wastes, development, to be sustainable, has to be contained within the constraints imposed by this environment. Planners, managers or developers must keep in mind that, although natural resources cost nothing to make and little to take, when destroyed they are difficult or impossible to replace.

This reaffirms that, as already indicated, development in any area should be kept within the ecologically sound 'solution space' identified for this area, during the planning process, in its territorial, quantitative, qualitative and temporal dimensions.

A tool for defining this space was developed by introducing a concept of *environmental thresholds* as specific types of thresholds that are imposed directly by natural resources (Kozlowski, 1986). Some of them indicate 'final' limits of what the natural environment can take without being irreversibly damaged, or destroyed. They have been called Ultimate Environmental Thresholds (UETs) (ibid.).

The 'threshold' concept has been known and applied in science for a long time and, more recently, some authors have used it in relation to ecologically sustainable development. An ecosystem, for instance, may sometimes 'cross a threshold into a new and usually unexpected condition, which itself may turn out to be in dynamic equilibrium under the new conditions' (Alpin *et al.*, 1995: 12). Barrow (1995) applies this concept extensively, in relation to various resources, arguing that decision makers need to recognise crisis points, or *critical environmental thresholds*, before they are reached, as then they can act and avoid problems.

Pearce and his colleagues (1989) warn about 'threshold uncertainty', which means that targeting a particular year for assessing impact could be often misleading, as it would not show when a change that really matters, would occur. For example, 'an impact that does not occur in 2030, say, cannot be assumed to be only marginally worse in 2035: it may be associated with the passing of a threshold' (Pearce *et al.*, 1989: 14). It seems to be generally recognised that thresholds in nature occur when consequences of some processes lead to sudden and sharp changes. Yet, 'though very important, thresholds for many events are not known or are poorly understood' (Pickering and Owen, 1994: 8).

To facilitate definition, or at least understanding of environmental thresholds, the *threshold approach* was introduced to planning and an original UET method was designed (Kozlowski, 1986). This method, it is believed, can make an effective practical contribution towards planning for sustainable development.

The UET has been defined as:

> the stress limits beyond which a given ecosystem becomes incapable of returning to its original condition and balance. Where these limits are exceeded, as a result of the functioning or development of particular activities, a chain reaction is generated leading towards irreversible environmental damage of the whole system or of its essential parts.
>
> (Kozlowski, 1986: 146)

The method has been based on the assumption that the four major types of UETs – territorial, quantitative, qualitative and temporal – can jointly define the desired 'solution spaces' for particular activities. It has also been assumed that this can be achieved through confronting potential environmental threats with the quality of respective environmental elements and through investigating their '*uniqueness*', degree of occurrence; '*transformation*', degree of change as compared with original state; '*resistance*', degree of ability to withstand negative impact and to self-regenerate; and '*biological importance*'.

Attempts to apply the method in planning practice began during the preparation of a regional plan for the Tatry National Park (TNP) in Poland (Baranowska-Janota and Kozlowski, 1981; Kozlowski, 1986) and were followed in Australia, in the Capricornia Section of the Great Barrier Reef Marine Park (Kozlowski, 1986; Rosier *et al.*, 1986; Hill *et al.*, 1987; Kozlowski *et al.*, 1988), in the Brisbane Forest Park (Charters, 1985; Roszkowska-Kuiken, 1987) and, more recently, on North Stradbroke Island (McNamara *et al.*, 1993; Alberti *et al.*, 1993).

The UET method primarily offers a 'way of thinking' which, in any specific situation, allows the definition of which environmental factors, where, why and how, limit particular developments. So far, however, its practical applications have concentrated on tourism development in environmentally sensitive areas (national, marine and forest parks). Nevertheless, from the very start it has been consistently argued that 'the approach can also be applied to other human activities and to other, not necessarily protected, areas' (Kozlowski, 1986: 183).

The method may become, therefore, a useful tool in the formulation of development strategies or policies and in planning studies that are set to find the best site for a specific development. In particular the method can be of assistance:

* *during the 'restrictive strand' in the preparation of planning schemes* where the main responsibility should be to minimise long-term negative effects of proposed developments on existing levels of environmental quality and to provide the stepping stone for making development sustainable; and

- *during the review of planning schemes* where the ecological reliability of land use allocation can be checked against 'territorial' UETs. This may be followed by indicating an ecologically acceptable scale of future development derived from 'quantitative' UETs.

It may be argued that there is a justified need to promote the method as an important element in the planning and decision-making processes at various levels. Experience, however, shows that difficulties often emerge when theoretical principles are put to practical tests. But such difficulties cannot be overcome without the tests being undertaken.

Two immediate practical problems, usually faced when trying to introduce and promote the method, must be disclosed:

- To be accepted by practising planners, the method needs to be fairly rudimentary. Yet, it cannot be applied without essential input from environmental scientists who may find it difficult to agree with what can be seen as ecological primitivism, promoted by the planners, and to accept the 'rules of the game' in the everyday reality of planning practice. These rules require accepting that plans are commonly produced within a fixed and usually low budget and within a fixed and usually too short time-frame. Thus, they prevent the in-depth analyses and empirical surveys required by any competent scientist who might be persuaded to co-operate with the planners.
- The method may, on the other hand, be considered by planners in the field as far too 'green' or academic and, therefore, of little use in the 'real world'. Many of them would take the view that the ecological content of their plans is already quite satisfactory. In fact it is often less than modest, even though practising planners tend to believe that a thorough *description*, and not *analysis*, of the natural environment is quite sufficient to formulate ecologically sound planning proposals!

Conclusions

Human health is inextricably linked to the environment and to sustainable development. This can be related directly to concerns resulting, for example, from chemical and biological contamination which have immediate and readily measurable effects on human health. In cases such as these, which may relate to an explicit issue such as disease, there is the opportunity to address the problem directly through, for example, medical technology and attention to a specific spot-source of contamination. While the application of such remedial action is achievable, its distribution across the economies of the developed and developing world is not. At a different level, it is unfortunate that the broader issues involved with wholesale destruction of ecosystems are not as easily treated as the simple cause–effect examples outlined above.

At the broader level, human health can be related indirectly to the environment through issues such as disruption of ecosystems which ultimately impinge upon the ability of the environment to produce food, shelter and spiritual succour. While the run down of ecosystems which elicits these effects may take long time spans to develop, the process is now well understood. Although this latter type of impact may have catastrophic results for individuals, societies, or whole nations, as in the case of famines, there is an apparent reluctance on the part of decision makers to address the root causes of such calamities.

To ensure a healthy environment for present and future generations, there must be an emphasis on ecologically sustainable development. Policies and practices which allow environmental damage and ultimately affect human health, are not sustainable. This is the only approach which will directly ensure the viability of the ecosystems upon which humans and all life ultimately depend. Human health is inextricably linked to the viability of the concept of ecologically sustainable development.

The status of the developing world influences the way in which its inhabitants perceive and interact with their environment. Without adequate food, shelter and *health* there is little option but to destroy the environment in a grim effort to survive (Tolba, 1987). From a global perspective, the greatest challenge facing the proponents of ecologically sustainable development is how to use the concept of sustainable development in helping the poor. The absence of sustainable health for the world's underprivileged is fundamental to these concepts.

It is now increasingly recognised that this problem is not restricted to developing countries. There is an appreciation and growing concern that sustainable development must be implemented, not only for humanitarian reasons, but also in order to protect the world environment as a whole. With a fierce urgency, both the developed and developing worlds are coming to the realisation that their futures are bound in common reliance on the global ecosystem.

Within this milieu of direct and indirect effects on human health, sustainable development and its relationship to ecosystems theory, there is a crucial role to be played by the planning profession. It does have its finger in the 'development pie' in most countries, although the levels of involvement and sophistication vary greatly. Given the unique position of planning as the 'organiser' of the multidisciplinary decision-making process which is currently gaining favour, planners have a responsibility, and opportunity, to advance the cause of sustainable development. And, as has been argued in this chapter, there is a direct link to sustainable health with implications for the local, national and global scales. In fact, the case could be put that the issue of *public health* relies very heavily on the role, influence and contribution of planning. Given the contemporary trend for equating sustainable development with a pairing of human welfare and ecological sustainability, now may be an appropriate time for planners to reassess their position and take a more proactive role in promoting ecologically sustainable development.

Acknowledgement

Some parts of the text, dealing with relationships between planning and sustainable development and with the presentation of the Ultimate Environmental

Threshold (UET) method, are taken from the first two chapters in *Towards Planning for Sustainable Development*, Jerzy Kozlowski and Greg Hill (edit), 1993, Ashbury, Aldershot.

References

Alpin, G., Mitchell, P., Cleugh, H., Pitman, A. and Rich, D., 1995, *Global Environmental Crisis*, Oxford University Press, Melbourne.

Alberti, M., Foster, T., Griffin, M., Hruza, K. and Outhwaite, S., 1993, North Stradbroke Island: a golf course location, in: Kozlowski, J. and Hill, G. (eds), *Towards Planning for Sustainable Development*, Ashbury, Aldershot, Brookfield USA, Singapore and Sydney.

Baranowska-Janota, M. and Kozlowski, J., 1981, *Method of Allowing the Identification of Ultimate Development Thresholds from the Viewpoint of the Protection of the Natural Environment*, Final Report, No. EPA 908/5–81–004A, Environmental Protection Agency, Washington DC.

Barbier, E., 1987, The concept of sustainable development, *Environmental Conservation*, Vol. 14, No. 2.

Barrow, C.J., 1995, *Developing the Environment*, Longman Scientific & Technical, London.

Carew-Reid, J., Prescott-Allen, R., Bass, S. and Dalal-Clayton, B., 1994, *Strategies for National Sustainable Development*, Earthscan Publications, London.

Chadwick, G.F., 1971, *A Systems View of Planning*, Pergamon, Oxford.

Charters, A., 1985, *Planning for People and Nature in Protected Areas*, Master's thesis, University of Queensland, St Lucia.

Chivian, E., 1993, Species Extinction and Biodiversity Loss: The Implications for Human Health, in: Chivian, E., McCally, M., Hu, H. and Haines, A. (eds), *Critical Condition – Human Health and the Environment*, MIT Press, Cambridge, MA and London.

Chivian, E., McCally, M., Hu, H. and Haines, A. (eds), 1993, *Critical Condition – Human Health and the Environment*, MIT Press, Cambridge, MA and London.

Christiani, D.C., 1993, Urban and Trans-boundary Air Pollution: Human Health Consequences, in: Chivian, E. *et al.* (ed.), *Critical Condition – Human Health and the Environment*, MIT Press, Cambridge, MA and London.

Cortese, A.D., 1993, Introduction: Human Health, Risk and the Environment, in: Chivian, E. *et al.* (eds), *Critical Condition – Human Health and the Environment*, MIT Press, Cambridge, MA and London.

Commonwealth of Australia, 1992, *National Strategy for Ecologically Sustainable Development*, AGPS, Canberra.

Coomer, J., 1979, The Nature of the Quest for a Sustainable Society, in: Coomer, J. (ed.), *Quest for a Sustainable Society*, Pergamon Press, Oxford.

Daly, H.E., Cobb, J.B., jr., 1989, *For the Common Good*, Beacon Press, Boston.

DHNAE, 1983, *National Conservation Strategy for Australia*, Department of Home National Affairs and Environment, Canberra.

Drewnowski, J., 1974, *On Measuring and Planning the Quality of Life*, Humanities, Paris.

El-Ashry, M.T., 1993, The World Bank post-Rio Strategy, *EPA Journal*, Vol. 19, No. 2.

Fowler, R., 1982, *Environmental Impact Assessment, Planning and Pollution Measures in Australia*, AGPS, Canberra.

Futures, 1988, Special issue on sustainable development, Vol. 20, No. 6.

Goodland, R. and Ledoc, G., 1987, Neoclassical Economics and Principles of Sustainable Development, *Ecological Modelling*, Vol. 38.

Gorzelak, G., 1986, Quality of life the regional perspective, in Kuklinski, A. (ed.), *Regional Studies in Poland*, Panstwone Wydawnictwo Naukone, Warszawa.

Hill, G., Kozlowski, J. and Rosier. J., 1987, A Threshold Based Reply to Tourist Development on Heron Island, *Queensland Planner*, Vol. 26, No. 1, pp. 10–15.

Holland, W.W. and Reid, D.D., 1965, The urban factor in chronic bronchitis, *Lancet*, No. 1.

IUCN, UNEP, WWF, 1980, *World Conservation Strategy*, IUCN, UNEP, WWF, Gland, Switzerland.

IUCN, UNEP, WWF, 1991, *Caring for the Earth – A Strategy for Sustainable Living*, IUCN, UNEP, WWF, Gland, Switzerland.

Kozlowski, J., 1986, *Threshold Approach in Urban, Regional and Environmental Planning*, University of Queensland Press, St Lucia, London and New York.

Kozlowski, J., 1990, Queensland Strategic Planning: Pitfalls and Prospects, *Queensland Planner*, Vol. 30, No. 2, pp. 3–7.

Kozlowski, J., 1993, Towards ecological re-orientation of professional planning, in Kozlowski, J. and Hill, G. (eds), *Towards Planning for Sustainable Development*, Ashbury, Aldershot, Brookfield, USA, Singapore and Sydney.

Kozlowski, J. and Hill, G. (eds), 1993, *Towards Planning for Sustainable Development*, Ashbury, Aldershot, Brookfield USA, Singapore and Sydney.

Kozlowski, J. and Hughes, J. (with Brown, R.), 1972, *Threshold Analysis – A Quantitative Planning Method*, Architectural Press, London and Halstead Press, New York.

Kozlowski, J., Rosier, J. and Hill, G., 1988, Ultimate Environmental Threshold Method in a Marine Environment (Great Barrier Reef Marine Park in Australia), *Landscape and Urban Planning*, Vol. 15.

Kuiken, M., 1990, Strategic Plans in Queensland: A Review of Gazetted Plans, *Queensland Planner*, Vol. 30, No. 1, pp. 31–36.

Leowski, J., 1986, Mortality from acute respiratory infections in children under 5 years of age; global estimates, *World Health Statistics Quarterly*, No. 39.

McNamara, K., Morris, C., Sorey, L. and Woodrow, M., 1993, North Stradbroke Island: a wilderness walk case study, in, Kozlowski, J. and Hill, G. (eds), *Towards Planning for Sustainable Development*, Ashbury, Aldershot, Brookfield USA, Singapore and Sydney.

Malisz, B., 1969, Implications of Threshold Theory, *Journal of the Town Planning Institute*, March.

Mazumdar, S., Scimmel, H. and Higgins, I.T., 1982, Relation of daily mortality to air pollution, *Archives of Environmental Health*, No. 37.

Munasinghe, M. and McNeely, J. (ed), 1994, *Protected Area Economics and Policy*, World Bank and IUCN, Washington DC.

Munn, R.E., 1988, *Towards sustainable development: an environmental perspective*, Paper presented at the Conference on Environment and Development, Milan, 24–26 March.

Niki, H., Dahy, E.E. and Weinstock, B., 1972, Mechanisms of smog reactions, *Advances in Chemistry*, No. 113.

Ostro, B.D., 1984, A search for a threshold in the relationship of air pollution to mortality: A reanalysis of data on London winters, *Environmental Health Perspectives*, No. 79.

Pearce, D., Markandya, A. and Barbier, E.B., 1989, *Blueprint for a Green Economy*, Earthscan Publications, London.

Pickering, T. and Owen, L.A., 1994, *An Introduction to Global Environmental Issues*, Routledge, London, New York.

La Trobe University and the Commission for the Future, 1989, *A Sustainable Healthy Future*, Recommendations of a national workshop, La Trobe University and the Commission for the Future, Melbourne.

Rodgers, K.P. and Saunier, R.E., 1994, Conservation in the Big Picture: Development Approaches for the Next Decade, in: Munasinghe, M. and McNeely, J. (eds), *Protected Area Economics and Policy*, World Bank, IUCN, Washington DC.

Rosier, J., Hill, G. and Kozlowski, J., 1986, Environmental Limitations: A Framework for and Development on Heron Island, Great Barrier Reef, *Journal of Environmental Management*, No. 223.

Roszkowska-Kuiken, M., 1987, *Adaptation of the Threshold Concept to Highway Planning in Valued and Sensitive Landscapes*, PhD thesis, University of Queensland, St Lucia.

Simonis, U.E., 1990, *Beyond Growth*, Edition Sigma, Bohn.

Thurston, G.D., 1989, Re-examination of London, England, mortality in relation to exposure to acid aerosols during 1963–71 winters, *Environmental Health Perspectives*, No. 58.

Tolba, M., 1987, *Sustainable Development – Constraints and Opportunities*, Butterworth, London.

UNEP, 1977, *Human settlements and habitat (Report of the Executive Director)*, United Nations Environment Program, Nairobi.

Wathern, P., 1988, Environmental Impact Assessment – Preventive Policy in Practice, in: Simonis, U. (ed.), *Preventive Umweltpolitik*, Campus Verlag, Frankfurt/M, New York, pp. 167–185.

Wilson, E.O., 1992, *The Diversity of Life*, Harvard University Press, Cambridge, MA.

World Commission on Environment and Development (WCED), 1987, *Our Common Future*, Oxford University Press, Oxford.

Zarsky, L., 1990, *Sustainable Development – Challenges for Australia*, Occasional Paper, Commission for the Future, Canberra.

6 Health and political ecology

Public opinion, political ideology, political parties, policies and the press

Nancy Milio

Introduction

The conditions for healthy living are shaped by public policies, mediated in their development and implementation by governmental and other organisations. Public policy is a guide to government action in order to alter what would otherwise occur; its purpose is to achieve a more desirable or at least acceptable state of affairs. An understanding of policy making can guide strategic planning so as to affect the direction and pace of policy making in order to improve the prospects for the health of populations; that is, it can increase the odds of designing and reaching health-supporting policy goals by guiding the selection and execution of the means to influence policy decisions.

Political ecology is here defined as those players and relationships that shape decisions about *who pays for* and *who gets what from* the determinants of health: education, employment, housing, health services, secure and healthy environments, access to information and to decision-making arenas. The players are the organised groups whose interests are affected by current and prospective policies. These include interest groups such as political parties, the media, bureaucracies, voluntary and commercial organisations. The media are in a unique position among organisations. They are not only conduits, creators and sculptors of information and issues; they are, as large profit-making corporations, interested parties in many types of public policies pertaining to taxation, workplaces, information and economic regulation. They are entrepreneurs and seek to attract audiences and advertisers (OTA, 1993; Cappella and Jamison, 1994).

All these players interact in a climate which arises from current and expected social, economic, political, technological and demographic conditions. That environment has already been influenced by past and current public policies. Thus both formal and non-formal policy-making activities are sensitive to the times and timing of actions undertaken to influence policy making, the timeliness or fit of proposed policies, the resources and strategic skills of the players as well as the concurrent accidents of history and nature (Rochefort and Cobb, 1993; Whiteman, 1985). These ongoing outcomes of policy making, among other things, shape the conditions for health of populations, especially disadvantaged sub-groups.

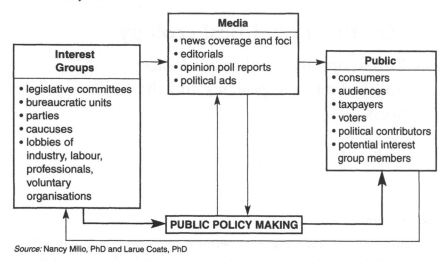

Source: Nancy Milio, PhD and Larue Coats, PhD

Figure 6.1 Primary policy-making processes affecting life and health of populations

As depicted in the schema (Figure 6.1), organised interest group activity primarily targets policy-making processes directly, and heavily uses the media to reach policy makers and the public indirectly. The public are those populations, most of whom are not active members of interest groups, who are affected by policy makers' decisions, which set the parameters for health in the form of access to services, products, prices and taxes.

The climate of policy making in the 1990s

The climate of policy making in the United States in the mid-1990s might best be viewed by looking at changes from the previous decade or more. As chronological benchmarks, the Democratic Clinton Administration took office in January 1993 and presented its long-awaited, comprehensive 1,500-page health care reform legislation in September. One year later, the bill was defeated and in November 1994 the Republican party won the majority in both House and Senate in the Congress, for the first time in 40 years. Its programme was a 'Contract with America' which sought to change government processes and size; to reduce taxes and eliminate the large federal deficit by cutting deeply into government spending on social but not military programmes, and fundamentally to reshape, decentralise, and privatise social, health and welfare programmes.

Public opinion

Public opinion polls taken prior to the election showed that, compared with the mid-1980s, some American values or beliefs remained the same, but others had changed. Americans were more racially polarised; they were more indifferent to the problems of the poor and minorities, and had grown more cynical about government competence (TMC, 1994).

Beliefs

Some beliefs had not changed; a majority retained faith in God (88 per cent), and prayed daily (78 per cent). They also believed that books with dangerous ideas ought to be banned from school libraries (51 per cent). A minority thought that women should return to their traditional roles (30 per cent), but more felt that too many children are raised in day care facilities, 75 per cent as compared with 68 per cent in 1987.

As regards civic responsibility the vast majority said that it was their duty always to vote (93 per cent) and expressed an interest in keeping up to date on national affairs (89 per cent). However, in the presidential elections in the 1980s, just over half of those eligible actually voted, and only 45 per cent voted in the 1994 Congressional Election that brought the Republicans to power. This victory was based on 23 per cent of the electorate. The poor and less well educated are less likely to vote (40 per cent) than their well-educated and affluent counterparts (over 60 per cent) (Bureau of the Census, 1995). These percentages may be compared with 89–95 per cent voting in elections in Australia, New Zealand, and Sweden, and 69 and 76 per cent in Canada and the UK (OTA, 1990; TMC, 1995).

Far fewer Americans (57 per cent) in the mid-1990s thought that government should care for people who cannot take care of themselves, down from 71 per cent in the mid-1980s. The figure for those who thought the poor were too dependent on government increased from 79 per cent in 1992 to 85 per cent in 1994. More whites (51 per cent) than blacks (26 per cent) thought that 'we have gone too far in pushing for equal rights' compared with the mid-1980s (46 per cent compared with 16 per cent). Fewer whites (44 per cent) than blacks (73 per cent) thought that there had not been much real improvement for blacks in the past few years (TMC, 1994).

Aid to the poor

Overall, fewer felt that the government should guarantee people enough to eat and a place to sleep, 59 per cent compared with 62 per cent in 1987.

A national random sample survey, taken just after the Republicans came to power in the 1994 election, divided respondents into those without and those with personal experience of receiving welfare payments at any time in their lives (Kaiser Family Foundation, 1995a).

Personal knowledge, as opposed to information gained though the media, made for some difference of responses. For example, when asked whether welfare payments should be ended after two years for those able to work, 70 per cent of those with no welfare experience, and 63 per cent of those with experience, agreed. Their answers were similar when the limit was applied even to those who could not get a job (18 per cent and 13 per cent); that is, the vast majority thought that payment should continue for people who could not find work. Most people in both groups felt that jobs were available (73 per cent and 70 per cent) but only 24 per cent and 18 per cent thought such jobs paid enough to support a family.

Majorities in each group believed that welfare aid encouraged women to have more children (71 per cent, 60 per cent), but a minority disagreed (21 per cent, 31 per cent), showing divergence in the two groups. A larger difference occurred when people were asked whether mothers should be denied more money if they had more children (64 per cent and 45 per cent). They disagreed again about whether welfare funds as a total programme should be limited ('capped') rather than as currently ('entitled') increased, whenever larger numbers of people needed assistance. Most of those who had never received payments wanted a 'cap' on spending (60 per cent) while 48 per cent of those who had received payments agreed with a cap.

Majorities in both groups favoured government job training and child care for women to enable them to leave welfare programmes (83 per cent and 90 per cent) and would pay more taxes for these programmes (54 per cent and 51 per cent). The severe welfare cuts and caps that were part of the Contract with America thus drew *selectively* on the opinion polls, e.g., few funds were proposed for job training and child care. (Kaiser Family Foundation, 1995b).

Government

People increasingly believed that elected officials lost touch with them when they went to Washington (83 per cent, from 73 per cent in the mid-1980s) and fewer believed that government benefits all the people (42 per cent from 57 per cent). More also believed that the Federal government 'controls too much of our lives' (69 per cent from 58 per cent) and that regulation of business does more harm than good (63 per cent compared with 58 per cent formerly). In ranking how favourably those asked view public institutions in 1994, the military came top, gaining 14 points from 1990, and business gained 15 points to come second, while Congress lost 21 points, with only 53 per cent of people giving it a favourable rating (TMC, 1994).

Political ideology

Political ideology may be thought of as beliefs that are not based on verifiable reality or are based on only very selected pieces of social reality. As such, they are often simplistic statements that can be used rhetorically to further political interests; they are useful political myths. The Contract with America contained an important component on changing 'welfare as we know it'. This entailed such reductions as a two-year limit on benefits; no increase in payments for women who bore additional children, and a lifetime limit of five years for any individual. These restrictions were said to be based on certain beliefs which, in fact, conflict with social data.

For example, Republican Party leaders held that welfare aid was producing the increase in single parent families (i.e., out-of-wedlock births). The data show that two-thirds of the rise in such families was due to an increase in single parents, in divorce, and in cohabitation without marriage. Other contributing factors were the rise of women in the workforce (58 per cent of those over 16) and their increased income; the drop in male earning power by almost 9 per cent; and the greater break-up rate of poor families. Finally, by comparison, other Western countries provide more family benefits but have lower rates of out-of-wedlock births and single parent families (Centre on Hunger, 1995).

A related ideological claim, used to justify the reduction of welfare benefits, was that if single parent families were a major cause of poverty, then welfare aid causes a rise in poverty and therefore should be severely restricted. Data show, however, that single parent families over a long period account for only 1–4 per cent of child poverty. The primary cause of poverty in the US is, rather, a lack of jobs at above-poverty wages. Low wages account for 85 per cent of child poverty (Centre on Hunger, 1995). Over the past twenty years the largest share of new jobs has been in the low wage sector, resulting in a drop of more than 19 per cent in real weekly wages and a growing gap in the distribution of income (*Economic Report*, 1994). As a result, between 1972 and 1992, the richest 20 per cent of the population increased their share of the national income by 3.3 per cent to 46.9 per cent, while the middle quintile declined by 2.7 per cent, and the share of the poorest fifth dropped from 4.1 per cent to 3.8 per cent (Bureau of Census, 1993).

Policy making based on such ideological underpinning results in adverse health and social effects on the social groups whose voice is not heard in the policy-making process, as suggested by a study on impact in the last section of this chapter.

Public opinion and public information

People base their beliefs and opinions, of necessity, on what they directly experience and what they learn through other channels of information, from other persons or from some form of media. All such conduits select and shape the information they pass on, according to their own priorities. Most Americans get most of their information or news through the mass media, primarily television, and this applies particularly to those on a lower income and those educated only to secondary level.

The mass media thus provide the basis on which public opinion is formed. When opinion polls are also reported by the media, the selection and shaping of the findings may affect public opinion, as suggested later. Public opinion is used to buttress political claims. How sound is such public opinion? How well founded on social facts is it? The quality of public opinion depends on the sources of information most people have access to, and also on how easily they can access a wide range of sources. These issues are further examined in a number of studies outlined below.

Public knowledge and public issues

In the late 1980s, surveys showed that although three out of four Americans were 'aware of political issues', up to four in ten could not answer more than one out of three questions about three high-profile national issues (OTA, 1990).

In the mid-1990s, a majority of Americans did not know the benefits they were entitled to under the national health insurance programme for the aged and disabled (Medicare); nor the causes of its rising costs, nor major programme components in the Federal budget as related to deficits – all issues having high political and press visibility. Yet these were issues on which the public was constantly being asked to voice its opinion (Kaiser Family Foundation, 1995a).

Information sources and uses

The vast majority of people used television and news magazines as their main sources of news, and between 60 and 70 per cent used the newspapers and tabloid TV, while almost half listened to talk radio. Between 25 and 33 per cent of Americans used public television or radio. Yet their confidence in these sources was not great: fewer than half thought television was credible, while only one in four believed the news on radio and in magazines (TMC, 1994).

Most people relied on the electronic media as their source of health information, regardless of their level of education, while more educated people were also more likely to use books (Harris & Associates, 1987). However, television was used mainly for entertainment. While half of the population read books, one in ten did so for general knowledge as opposed to entertainment and 'how to do it' reading. Figures for magazine buyers were similar (OTA, 1990).

Diversity of sources

Access to new and more diverse sources of information and points of view was related to income. In the mid-1990s, while television was equally prevalent among low and high income groups, the gap increased by 20 or more percentage points for newer technologies such as home-based cable TV and VCRs, as well as telephones. Far larger gaps occurred for computer-based technologies (Interactive Services, 1994). These patterns were similar in the UK (McKinsey, 1993).

A national survey of access to networked information technology, i.e., computers linked to electronic networks, revealed that among people with less than $10,000 yearly income, fewer than one in ten had home PCs, and of these only one or two (i.e., 2–4 per cent of the poor) were networked (NTIA, 1995). This compared with about two-thirds of high income people (with over $75,000 income) who had home PCs, of whom about six in ten were networked, or about 40 per cent overall. Rural populations also had less access, generally less than urban groups.

Information sources and policy making

The significance of information access, especially access to the new information technology, for policy making, concerns access both to timely and accurate information and to participation in policy-making processes through coordinated action. An unexpected finding of the above survey was that among all those with a networked computer, the network was used equally by high and low income groups for job searches and for taking online training courses, with those on a lower income using educational services somewhat more often. Further, accessing online government reports was done equally by both income groups, at a rate of 10–20 per cent, and more so among minorities than whites (NTIA, 1995).

If these early uses of electronic networks by disadvantaged groups herald similar interest by their peers, then easy access to such technologies in the libraries, schools, and other community facilities in low income areas could importantly increase their participation in the policy-making processes that affect their conditions for health and life. This prospect then has implications for equity in policies involving development of information technology infrastructure and support services.

Another new policy-making dimension of the new technologies is occurring at national levels, where the political arena has widened vastly through interactive TV via cable, satellite, and multimedia computers on broadband networks. It is often in this electronic theatre that new issues are raised, solutions proposed, and legitimacy of individuals and groups established. Issues in these mass media, however, may be too simply defined, solutions too superficial, and legitimacy conferred on only selected groups because of the criteria of the 'market' – what 'sells' to audiences and advertisers (OTA, 1990; Ladd, 1989).

This marketplace aura has resulted in the rise of political consultants whose tools are marketing research and political advertising. In a sense, although the policy arena has expanded, the price of entry set by the new gate-keepers has increased. In the US, for example, the amount spent on political campaign advertisements has grown over 200-fold since 1970 to well over a quarter of a billion dollars in the early 1990s (OTA, 1990).

The press

The press, as mass print and electronic media, are thus not neutral conduits in the policy-making processes that affect the prospects for people's health. They are agenda-setters, shapers of opinion, and virtual legitimisers and convenors of public groups, of political opponents and allies. They are also businesses with interests in operating at a profit by selling audiences as markets to paying sponsors and are concerned with the policies and politics that affect their interests.

News coverage

A picture of these kinds of actions and impacts in the US mass media may be formed from the coverage of the 1993 health care reform debate during the first half of the term of the newly elected Clinton Administration. A major study of the media was based on a content analysis of 17 leading national and regional newspapers, magazines, TV and cable networks. They were covering the introduction of the reform in September 1993 and tracking its progress to the November 1994 election of a Republican Congress following defeat of the reform (Braun, 1995).

Tracking both public opinion and the extent of media coverage of the debate, the study showed that public support peaked at 59 per cent when the reform was introduced and began to drop in January 1994, falling to a low of 40 per cent in July when the bill was mired in Congress. It was defeated in October 1994 as pre-election rhetoric intensified and the Congressional clock ran out at adjournment.

Throughout this period, the media followed the political process, focusing on Congress, not the Clinton Administration or the issues, and did not expand the coverage in relation to the public's high interest. If it had, the coverage would have been broader and more sustained, focusing more on substance than political jockeying. Polls showed that public knowledge of the reform dropped over time as coverage fell; public dissatisfaction with media coverage at the same time increased. For example in September 1993, 44 per cent of people said that media coverage was good or excellent, but by August 1994, 32 per cent rated it thus. In effect, the media focused on the political game, not the impact on people and their health of this historically unique legislation.

Of the more than 2,000 stories published during the 15 months, television, the medium which most people used for news, carried only 16 per cent. The tenor of all the stories became less neutral over time as regards the prospects for adoption of the legislation, changing from a high of two-thirds neutral in the beginning to 20 per cent neutrality towards the end. Negative slants increasingly outnumbered positive ones about political prospects for the reform package.

Political advertising and the news

Another facet of the impact of the media on the political process is the increasing coverage by news programmes of paid advertisements sponsored by interest groups. Detailed studies showed that about $50 million was spent by interest groups to influence the result of the health care reform debate, most of it on media advertising. The purpose was to influence reporters and Congress, not the general public, and most of the broadcast advertisements (59 per cent) were judged to be unfair, misleading, or false. But newscasters' coverage of the material focused not on its fairness or accuracy but on the strategic purposes of the groups (Annenberg Public Policy Centre, 1994b).

Such coverage dramatically increased the audience and in the process legitimised the sponsors as serious players in the debate. In turn, the makers of the advertisements increased their emphasis on attacking their opponents, hoping to garner news coverage. As a result, the television time showing the material was expanded by a total of almost 15 minutes of free time (unpaid by advertisers) over the course of a year, roughly equivalent to almost $1 million in paid advertising time (Annenberg Public Policy Centre, 1994a).

Media focus and style, and public attitudes

Television national news reports since the late 1980s focused increasingly on strategy, tactics, and opinion polls, rather than coverage of policy issues. Issue coverage dropped to one-third of political news, while two-thirds was given to strategy and polling reportage. Similar trends were apparent in even the most highly regarded newspapers. Studies of the effects of this focus and style of reporting on public attitudes were made during the health care reform debate, showing that they induce cynicism in audiences and undermine confidence in government (Capella and Jamison, 1994).

Cynicism increased in readers, and especially in TV audiences, when the media focused on strategic 'angles' such as who or what is winning or losing; the viewing of policy positions as tactics to gain support rather than securing it by the use of principled statements; the reporting on slogans or political advertisements and polls rather than the analyses of issues in audience-friendly ways, such as personalised stories and simple descriptions.

Even when substantive issues were presented, they were posed in oppositional terms. Each 'solution' to a problem was countered by a contradictory one, leaving the audience with the impression that there was no solution, especially of complex issues, such as health care or welfare reform. Simple narratives and 'personal stories' may be less conducive to cynicism, but may not be sufficient to educate an uninformed audience in helping them to decide between alternatives (Cappella and Jamison, 1994).

News slant and public information

An example of how the choice of focus or 'slant' of a health policy news story can affect the way in which the media help or fail to create an informed public, and therefore public opinion, occurred in September 1991 (Moore, 1992).

The major TV networks have paired themselves with leading national newspapers to conduct frequent public opinion polls. One such poll on health insurance benefits was carried out by a combined *CBS–New York Times* polling group. The difference in interpretation between these two news organisations, which might not be noticed by most viewers and readers, is significant in illustrating the differences in 'message' that the public is offered.

CBS TV chose to focus its report of the poll on the personal lives of those who could not afford health care: why people had not gone to a doctor, for example, and what that meant for their individual well-being. Alternatively, the *Times*, reporting the same poll, carried a front page article that headlined a different slant from the CBS-TV News story: 'Health Benefits Found to Deter Job Switching', accompanied by a graphic titled: 'Job Lock'. This was a series of pie charts showing the percentage of adults who said that because they did not want to lose their health insurance coverage, they had stayed in a job even though they wanted to leave it. The pattern showed that this 'job lock' was much more likely to occur in the two middle income groups than in the highest or lowest income groups.

The *Times* had considered at least four possible slants to the story. Three were rejected because they were not especially new, having been reported in previous polls. These included the findings that nine out of ten people saw a need for a fundamental change in the US health care system; that eight out of ten said the country was heading towards a health care crisis; and that three out of ten had no health insurance at some time during the previous year. All of these points were covered in the story, but were not the lead. One finding that had not been previously reported was the number of people who were in 'job lock' due to their fear of losing health insurance. The *Times* decided that this was particularly newsworthy because middle income people were the most affected. And it was the middle class that the Democrats were targeting in their effort to regain the White House.

Thus, although based on the same poll, the two reports from CBS and the *Times* differed considerably. The television network's story was driven by the more intimate nature of the medium, which could easily portray the emotional consequences of not having health insurance and not having health care. The newspaper's story was shaped by the more abstract nature of the print medium and its view of the political importance of its finding to its readership. About three times as many poll questions were analysed in the *Times* story as in the CBS News broadcast, and the results were illustrated with charts and graphs in a lengthy, major story. Each medium selected the content and format that it saw as most attractive and compelling for its own audience.

Many academics argue that the media actually make news when they conduct and report their own polls, rather than simply report news. In response, the media's pollsters point to the importance of public opinion in American government and insist they should report it on a regular basis, much as they report news regularly about other activities of government. While polling can provide a continuous monitoring of public viewpoints, the tendency for media polls to ask forced-choice questions on some topics that are unfamiliar to most people, can also create the *illusion* of public opinion (Moore, 1992).

The commercial imperative and health issues

The commercial interests of the media, as large corporate enterprises, often result in the polarisation of issues rather than a reasoned discussion or analysis. The impact of this practice on public knowledge and opinion stems from the fact that the American public gets most of its news about health from television and newspapers (Nimmo and Combs, 1985). In theory, the media have the 'reasonable reader' standard, which requires that information about the risks and consequences of what is being reported be in a form that is accurate, understandable, and substantially complete. But the mass media need stories that are 'new and different', timely, close to their audiences, having 'human interest', simplicity, and controversy to meet the imperatives of the journalistic market (Agee *et al.*, 1983; Klaidman, 1990).

Long term, 'non-event' stories pose problems of time and money for the media. For example, the lethally polluted village of Bhopal in India was an emblematic community–corporate–government environmental conflict. It contained all that the media required as a 'good story': novelty, timeliness, controversial health risk information, people voicing outrage, and corporate and government officials as blameworthy (Greenberg and Wartenberg, 1990). On the other hand, the lengthy time process and the smouldering hazards of Bhopal prior to the crisis did not meet media criteria for a good story and so were never reported.

Other examples of health-related 'non-events' are the AIDS epidemic before it was manifested in celebrities, nuclear power plants before Three Mile Island, and cigarette smoking prior to the US Surgeon General's elaborately planned media event of 1964. The scientific media had reported, as far back as 1938, evidence of a close relationship between smoking and disease. But there was strong resistance by the press to publishing this news because more than half of adults smoked. Cigarettes were a multi-billion-dollar 'All American' industry during a world war and were advertised heavily in the media; and journalists smoked. The media emphasised industry-sponsored research and under-played the health effects of cigarettes (Klaidman, 1990). The result was that the story was not visible in the mass news media, and so not part of the education of the public nor supportive of an informed citizenry that would eventually have to indicate its views, in polls or votes, for policy positions to address the issues.

Even when the mass media report health-related stories, the commercial imperative which requires audience-grabbing style and slant, emphasising treatment rather than prevention, is often piecemeal, presented in unclear ways, and ignores the underpinning science, tending to focus instead on one or two accessible articles (Houn *et al.*, 1995; Freimuth *et al.*, 1984; Angier, 1993). This kind of reporting, and therefore the 'education' of the mass media-oriented public, affects public understanding of policy issues and what to do about them (Neuendorf, 1990). Vast amounts of information are available. Yet, its delivery by the most widely used media is not creating an informed public able directly to influence policies that affect well being and health. Nor do the prevalent media channels inspire trust in either government or the media.

In effect, the new information technology, the mass media, polling tactics, and interest groups have developed into a sophisticated system to target ready audiences, potential sources of supportive blocs of votes and constituencies for ideological mobilisation. By combining survey, census, and electoral databases, interest groups can target small local areas, shape their 'message', usually in 'sound-bite' form, and make contacts with individuals through telephone banks,

door-to-door canvassing, direct mail, or local cable TV advertising and program-
ming (Kramer and Schneider, 1985). These techniques can then strengthen
extreme factions within political parties by giving effective voice to their ideological positions, whether or not those positions are congruent with *verifiable* social
reality!

Political parties and other stakeholders

The case of health care reform

Given the widespread public sense in the US of less collective responsibility, of
growing social divisions, and of scepticism about government – all influenced
at least partly by the style and information focus of the mass media and interest
group techniques – the 1993 health care reform proposal energised
conservatives to accelerate well-honed electronic media skills and raised
public consciousness about the role of government. The result was that those
who stood to gain most from the reform were less mobilised than the
opposition to reform, and Democratic party leaders did not push the reform
package because they feared that, given public opinion, their re-election was at
risk if they identified too closely with the 'big government' President (TMC,
1995).

In a study of the defeat of the health care reform proposal, 56 major party leaders and their legislative staff were asked their views (Columbia Institute, 1995).
They generally agreed that the policy itself was too complex and confusing, that
the policy-making process had not involved Congressional party leaders early
enough, and that there was too long a delay between introduction of the legislation and the final vote. This long period allowed the opposition to mobilise and
grow. Such reasons, of course, may be true for many bills; further, since the
Democrats were in control of both the House of Representatives and the Senate,
the time frame was in their hands.

Perhaps more telling was the pervasive view among these leaders that public
opinion was as influential in the debate as the Clinton Administration itself; that
interest group advertising, mainly by the health insurance and business interests,
was important; that the public was not well informed on the issues, and that the
media had done a poor job of helping people to understand the issues. Despite
these views, the informants also said that their main sources of information about
public opinion were the polls, which were mainly reported in the media and by the
trade group lobbies.

These findings further reveal the conduit role of the media, but also suggest
how the media, through style and slant embedded in their own commercial
imperatives, help to shape public opinion, and in turn impact upon policy making.
Policies, or their absence, eventually flow back to impact on conditions for healthy
living of the public.

Policy impacts on health

Policy substance

Under Federal legislation prior to the Contract with America, welfare support was far less generous than in other affluent nations. The 1995 poverty line was set at $1,073 per month for a family of four, while the minimum wage amounted to $737, and the average benefits for families receiving assistance across the 50 states, which can supplement benefits, was less than $800. This amount included cash, housing, food, and medical care benefits. Families with low-income wage-earners who received such benefits had a monthly total of $950. (GAO, 1995).

Against this background, in a preview of what might occur as Congressional welfare and health policy changes are implemented at the level of the individual states, one, Michigan, initiated a 'reform' of its welfare programme in 1993, based on the conservative ideology described above. The Michigan welfare pro-gramme had granted a small monthly cash allowance, free medical care and food stamps to those who did not qualify for any other welfare programme. The reform ended the programme completely. One of the beliefs of policy makers was that the programme supported young black males who were able to work, and who, therefore, would now be forced to find jobs. In fact, only 1 in 20 beneficiaries fell into that category; 40 per cent were women and 40 per cent were over 40 years of age.

A follow-up study was made of a random sample of the former beneficiaries two years after the end of the programme (Daninger and Kassoudji, 1995).

After two years, 20 per cent of the entire sample were employed, while about 13 per cent of the total had never had a job before. Most of the jobs were janitorial or kitchen work at an average $5.56 per hour at 35 hours per week. The estimated hourly rate to be able to afford low cost rental housing is over $13. Thus, with this increase of people at risk for affordable housing, the capacity of emergency urban homeless shelters was outstripped, affecting homeless single adults most.

About 35 per cent of the sample were 'doing as well' as when they were bene-ficiaries of the welfare programme, mainly through use of many unstable and small sources of funds, such as begging and selling their blood, while another 15 per cent were sufficiently disabled to qualify for other income support programmes.

Half of the total were in poorer health than two years earlier; over half suffered from or were at risk of clinical depression as compared with the usual finding of 25 per cent among jobless blacks. Six in ten had at least one major chronic disease. Half had some form of public or private care insurance, but fewer than half of those who worked had any.

Policy form

Not only does the substance of policy affect the conditions under which people can be healthy, such as income maintenance and housing and health services, but the form is also important.

A significant change in the form of Federal policy support was also a major part of the Contract with America. Rather than the traditional dispersal of funds to the states through matching grants for specific programmes, such as welfare and medical care for the poor and food stamps, the new approach called for bloc grants, previously available but only for minor amounts of funding. The use of bloc grants meant that there would be few Federal requirements about how the funds should be spent, who would be eligible for which benefits in what amount; thus potentially excluding some social groups which had previously received services. Also the states would receive over the next several years a smaller total amount of Federal funds, regardless of need.

Bloc grants would produce fewer services, unless states added new monies to the programmes, and this was not likely under budgetary conservatism. Further, there would be competition for available funds since there would no longer be guarantees that each social group would get its 'share'. For example, medical benefits for the poor might go to the elderly poor for nursing home care rather than to poor single mothers and their children for preventive and primary health care. The outcome was likely to depend on the political voice that could be raised for the affected beneficiaries (SEIU, 1995). The prospects for the health of the most vulnerable groups would be grave.

Conclusion

The political ecology of policy players, their web of relations, and their increasing use of old and new electronic media as multi-purpose tools for influence and action, pose both threat and opportunity to those who are concerned about the impact of policy on health. The complex connections between the players in the policy arena suggest that the substance, form, and process of policy making are critical to the impacts of public policy on the options and prospects for the health of populations. An understanding of the processes and the interests and strategies of the major players can help to inform approaches which, applied in particular local or national contexts, can influence policy decisions to bring about more healthful impacts than might otherwise be.

References

Agee, W., Ault, P. and Emery, E., 1983, *Reporting and writing the news*, Harper & Row, New York.

Angier, N., 1993, Science and the press: From star-struck lovers to uneasy bedfellows, *American Society of Preventive Oncology Annual Conference*, March 21, Tuscon, AZ.

Annenberg Public Policy Centre, 1994a, *Newspaper and Television Coverage of the Health Care Reform Debate*, Research report, 16 January-25 July, Philadelphia, PA.

Annenberg Public Policy Centre, 1994b, *The Role of Advertising in the Health Care Reform Debate*, Research report, July 18, Philadelphia, PA.

Braun, S., 1995, Media coverage of health care reform: A content analysis, *Columbia Journal Review*, March/April, Supplement: 1–8.

Bureau of the Census, 1993, *Money Income of Households, Families, and Persons in the US,*

1992, September, Department of Commerce, Washington DC.

Bureau of the Census, 1995, *Turnout in 1994 Congressional Election*: 45 per cent; Younger Voter Participation Shows No Gain, Press release, 8 June, Department of Commerce, Washington DC.

Cappella, J. and Jamison, K., 1994, *Public Cynicism and News Coverage in Campaigns and Policy Debates: Three Field Experiments*, Research report, 4 September, Annenberg School of Communication, University of Pennsylvania, Philadelphia, PA.

Centre on Hunger, Poverty and Nutrition Policy, 1995, *Statement on Key Welfare Reform Issues: The Empirical Evidence*, Tufts University, Medford, MA.

Columbia Institute, 1995, *What shapes law-maker' views? A survey of members of Congress and key staff on health care reform*, May, Washington DC.

Daninger, S. and Kassoudji, S., 1995, *When Welfare Ends: Subsistence Strategies of Former General Assistance Recipients*, Research report, February, University of Michigan School of Social Work, Ann Arbor, MI.

Economic Report of the President, 1994, Government Printing Office, Washington DC.

Freimuth, V., Greenberg, R. and DeWitt, J., 1984, Covering cancer: Newspapers and the public interest, *Journal of Communication*, 34: 62–73.

General Accounting Office (GAO), 1995, *Entitlement Programs*, US Congress, Washington DC.

Greenberg, M., Wartenberg, I., 1990, Risk perception: Understanding mass media coverage of disease clusters, *American Journal of Epidemiology*, 132, Supplement 1, S192–194.

Houn, F., Bober, M., Huerta, E., *et al.*, 1995, The Association between alcohol and breast cancer: Popular press coverage of research, *American Journal of Public Health*, 85: 1082–86.

Interactive Services, 1994, *A Survey of Privacy and American Business*, Interactive Services, Washington DC.

Kaiser Family Foundation, 1995a, *How Americans With and Without Welfare Experience View the Welfare System and Welfare Reform Proposals*, Kaiser Family Foundation, Menlo Park, CA.

Kaiser Family Foundation, 1995b, *New Survey Finds Most Americans Oppose Slowing the Growth of Medicare*, Press release, 29 June, Kaiser Family Foundation, Menlo Park, CA.

Klaidman, S., 1990, How well the media report health risk, *Daedalus*, Fall, 119–132.

Kramer, K. and Schneider, E., 1985, Innovations in Campaign Research, Finding the Voters in the 1980s, in: Meadow, R. (ed.), *New Communications Technologies in Politics*, Annenberg Washington Program, Washington DC, 20–35.

Ladd, E., 1989, *The American Polity: The People and Their Government*, Norton Press, New York, NY.

Louis Harris and Associates, 1987, *Questionable Treatments, A National Survey*, Louis Harris and Associates, Washington DC.

McKinsey, R., 1993, *Public Service Broadcasters around the World*, Report, April, British Broadcasting Corporation, London.

Milio, N., 1992, Making Healthy Public Policies: Developing the Science by Learning the Art, An Ecological Framework for Policy Studies, in: Badura, B. and Kickbusch, I., *Health Promotion Research*, Oxford University Press, Oxford.

Moore, I., 1992, *The Super-Pollsters: How They Measure and Manipulate Public Opinion in America*, Four Walls Eight Windows, New York, NY.

National Telecommunications and Information Administration (NTIA), 1995, *Falling through the Net*, Department of Commerce, Washington DC.

Neuendorf, K.A., 1990, Health images in the mass media, in: Ray, E.B. and Donohew, L. (eds), *Communication and Health: Systems and Applications*, Lawrence Erlbaum, Hillsdale, NJ, pp. 111–135.

Nimmo, I. and Combs, J., 1985, *Nightly Horrors: Crisis Coverage in Television Network News*, University of Tennessee Press, Knoxville, TN.

Office of Technology Assessment (OTA), 1990, *Critical Connections*, US Congress, Washington DC.

Office of Technology Assessment (OTA), 1993, *Researching Health Risks*, November, US Congress, Washington DC.

Rochefort, I. and Cobb, R., 1993, Problem definition, agenda access, and policy choice, *Policy Studies Journal*, 21(1): 56–71.

Service Employees International Union (SEIU), 1995, *Block Grants: A State-By-State Analysis of the Fiscal Impact of Programme Consolidation*, SEIU, Washington DC.

Times-Mirror Centre for the People and the Press (TMC), 1994, *The New Political Landscape*, October, TMC, New York.

Times-Mirror Centre for the People and the Press (TMC), 1995, *The New American Electorate and Health Care Reform*, TMC, New York.

Whiteman, D., 1985, The fate of policy analysis in Congressional decision making, *West Political Quarterly*, 38(2): 294–311.

Part II

Health in micro ecosystems

7   Health of women

Changing lifestyles and reproductive health

Cristina Bernis

Abstract

The transformation of ecosystems implies changes in all the structural components, including the actual biology of the populations, a frequently forgotten fact
which deducts interpretative quality from the results of any health diagnosis of a
population. Changes in biology of women can result directly from nutritional
changes (affecting menarche, height, and so on), or from behavioural changes
related to the patterns of reproduction.

The comparison between Moroccan and Spanish women is used to evaluate the
extent and the consequences of changing reproductive patterns, which among
Western women have coincided with a progressive medical intervention in the
biology of reproduction. As evolutionary ecologists, we may wonder how much
of this medicalisation is necessary, because it is difficult to imagine that such an
important biological function as reproduction does not have regulatory mechanisms to deal with environmental change. The possibility that the degree of
environmental change could have affected the reproductive biology of women in
such a way as to result in large mal-adaptations is suggested, as it must not be forgotten that the biology of reproduction was selected in biological and
environmental circumstances completely different from the ones now prevailing in
Western populations. If the degree of environmental change has had this adverse
effect, it is time to analyse the problem globally and discuss possible solutions,
monitoring the biological and behavioural proximates of fertility, which could
reduce the risk of suffering the dysfunctions and symptoms mentioned.

Introduction

Since the publication of the Lalonde report in 1974 on the health of Canadians,
increasing efforts have been made to locate an understanding of health and disease
in an ecological frame of reference. In response to this, the WHO launched an
ambitious programme with two major objectives: the reduction of inequalities in
health, both within and between populations, and the promotion of healthy
lifestyles. Regional targets and indicators by which to measure their progress were
defined (WHO, 1985). The priorities for necessary research based on these goals

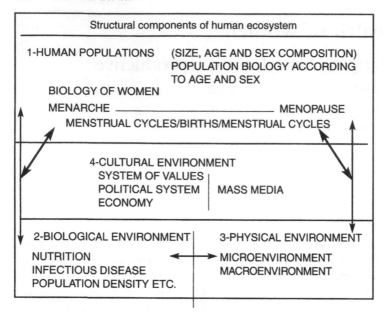

Figure 7.1 Structural components of human ecosystems

have been established, and the need for good systematic records and coordinated interdisciplinary research has been confirmed (WHO, 1988). Both human ecologists and human biologists have the evolutionary perspective needed to contribute to the theoretical framework and the methodological skills required for developing multidisciplinary research; with such means available we have the responsibility to do the work.

The ecosystem can be described as the basic unit of analysis, that global situation to which populations adjust. These adjustments are the physiological and behavioural responses of individuals and populations to environmental stimuli as a whole. Human ecosystems have four basic structural components: human population, biological environment, physical environment, and cultural environment. The cultural environment is exclusive to our species and fundamentally conditions the other three (Figure 7.1). The functioning of ecosystems is determined by the continuous interchange of energy among their structural components. This circulates in two cycles, one of them trophic, concerned with that vital energy which maintains us as biological units. The other cycle is composed of cultural or extrasomatic energy which maintains us as social units (Bernis, 1991b). The transformation of ecosystems implies changes in all these structural components, including the actual biology of populations. This is a frequently forgotten fact which detracts from the interpretative quality which unnecessarily limits or diminishes the results of any health diagnosis of a population.

Changes in the biology of women can result directly from nutritional variations influencing menarche, body mass index, height and type of menstrual cycle.

Behavioural changes may also result in reduction of family size, and avoidance or shortening of the duration of breast-feeding. At the same time, a general improvement of environmental and social factors may result in an extended life span (Bernis, 1991a). Important secular changes are taking place in the biology of Western women, and such biological change is expected to occur in women from other populations as modernisation and economic development take place.

When assessing health and disease patterns it is often difficult to distinguish the extremes of 'normal' biological responses to environmental changes from dysfunctional and (or) 'pathological' responses. In current environmental circumstances, it is therefore important to notice the extent to which any changes in normal biological processes can deviate towards a higher prevalence of serious dysfunctions and pathologies later in life.

Health and reproductive health

Health in an ecological context is that equilibrium between biological processes and the constraints of environment, on one hand, and between the individual needs and aspirations and the possibility that the environment can satisfy them, on the other hand.

The development of new environmental conditions in urban ecosystems results in new patterns of health and disease. This was summarised in the so-called 'epidemiologic transition theory' by Omran (1971). Later, Rogers and Hackenberg (1987) pointed out that for developed countries a new stage of epidemiologic transition has emerged, from the hybristic stage, into a new phase where health and disease outcomes depend principally on prevention strategies and changes in individual behaviour and lifestyles. In developing countries, however, infectious disease pattern continue to predominate, although there are wide differences in the tempo of transition towards new patterns of disease.

The term reproductive health has been applied to those reproductive events or gynaecological health problems occurring during the fertile life phase of women. According to this concept, reproductive health care is defined as 'that health approach which helps clients to reach their fertility goals in a healthful manner, by ensuring high quality family planning services and the offering of additional health services in order to help women with their multiple health needs at the same time and place' (Jain and Bruce, 1993).

This concept of service has been directed mainly towards women in the Third World. We must not forget that, until the 1980s, the main purpose of such services was to strengthen the access of women to birth control in order to curb population growth. Women's broader reproductive health needs were not of concern or priority (Ahlberg, 1991; Holloway, 1994). To alter this approach, it has been suggested that a wider perspective is needed, using the so-called 'life-cycle approach to health care' which takes into consideration all facets of women's experience and the way in which that relates to health (Holloway, 1994).

Exemplifying this perspective, the International Conference on Population and Development (held at Cairo in 1994) recognised the importance of reproductive

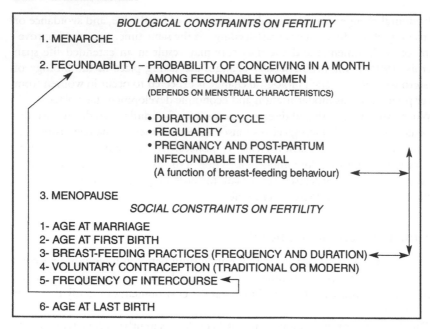

Figure 7.2 Biological and social constraints on fertility

health, and popularised the term through the media, 'as a reflection of health in childhood and adolescence, and also reproductive health sets the stage for health beyond reproductive years'. An ecological approach was also introduced, when it was stated that 'reproductive health affects and is affected by other aspects of health, nutrition, lifestyle and the environment'. This underlines the importance for reproductive life of health and nutrition during infancy, childhood and adolescence. The variability seen in patterns of reproduction is perhaps the best example of interaction between biology and culture. This allows the evaluation of the relative role of the physiological and behavioural mechanisms of population adjustment.

Enlarging our understanding of fertility as a major mechanism of biological evolution is of intrinsic interest. To this we can add the complexities of the social, political and sanitary consequences and how they affect the reproductive models in human groups (Henry, 1979; Bongaarts and Potter, 1983; Smith, 1983; Donaldson, 1991). The fertility of women is constrained by biological determinants, such as menarche, menstrual patterns and menopause, which characterise homo sapiens. As specific characteristics, they are the result of our species' evolutionary processes, and seem to have been selected to allow enormous flexibility with regard to the ecological and social milieu in which women experience life (Figure 7.2).

Reproductive biology in human groups depends on socio-cultural factors that act in two different ways. One is passive, through the action of ecological factors that regulate energy balance. The other is active, through socially established norms and values that regulate behaviours.

Those ecological factors, such as nutritional intake and energy expenditure, shape the variability in these biological determinants. When the balance is positive, menarche is earlier and menopause is delayed, thus prolonging reproductive capacity and bringing about competent menstrual cycles, 100 per cent ovulatory, at an earlier age.

Those social behaviours that regulate reproduction determine marriage age, ideal family size, duration and pattern of breast-feeding, and the patterns of contraceptive use in each society. Thus social behaviours shape the use of women's fertile life cycle without affecting its duration.

Modern urban ecosystems in Morocco are rapidly changing and show important parallels with a similar transformation that took place in Spain between twenty and forty years ago. Such past experience of ecosystems transformation could be used as a reference to evaluate the consequences that changes occurring in reproductive patterns in Morocco may have for the biology and health of women, and especially for those factors associated with a reduction in family size.

Secular changes in the biological determinants of fertile life: menarche, menstrual cycles and menopause

A secular decrease in menarcheal age has been described for Spanish populations, though differences are maintained between rural and urban women, and between women of different economic levels (Bernis, 1973; Bernis, 1980; Prado, 1984; Garcia-Moro and Hernandez, 1990). For Moroccan women and for women from other Arab populations, there is no such temporal decrease in menarcheal age (Bernis, 1991a; Prado, 1992). The reduction in mean or median age at menarche implies an increase in the number of women having their first period at 12 years of age or earlier. This is of interest because it has been shown that the pattern of cycle maturation, adult hormone levels, and therefore the existence of regular and ovulatory cycles, occurs earlier and more frequently in adolescents with early menarche than in those with late menarche (Vihko and Apter, 1984).

A highly significant association between early menarche and some menstrual dysfunctions, such as menstrual pain and pre-menstrual symptoms, has been described (Anderch and Milsom, 1982; Bernis *et al.*, 1994; Bernis, 1995). This fact might have far-reaching consequences for women's health, as we must not forget that the most frequent diagnosis associated with hysterectomies in the US, Canada and the UK is that of 'menstrual disorder' (Evans, 1979; Teo, 1990). A greater risk of breast cancer has been suggested according to the type of menstrual cycle and with early menarche (Apter and Vihko, 1983; Baanders and de Ward, 1992).

The use of birth control determines that a contemporary Western woman experiences many more menstrual cycles than her counterpart from a high fertility population. Figure 7.3 compares the percentage of menarche at 12 years of age or earlier in different Spanish and Moroccan populations.

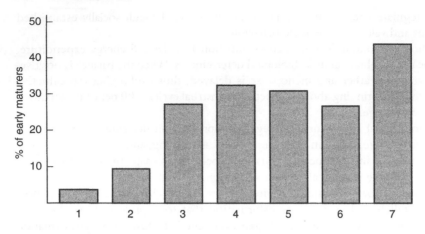

Figure 7.3 Prevalence of early maturers, experiencing menarche at age 12 or earlier, in selected Moroccan and Spanish populations

Sources
Col. 1: Marrakesh, Morocco, women aged 18–60: Verea, 1984
Col. 2: León, Spain, women aged 60–90: Bernis 1974
Col. 3: León, Spain, women aged 18–47: Bernis, 1974
Col. 4: Madrid, Spain, women aged 45–65: Bernis, 1994
Col. 5: Barcelona, Spain, women aged 45–70: Garcia-Moro and Hernandez, 1989
Col. 6: Marrakesh, Morocco, girls aged 13–18: Bernis *et al.*, 1996
Col. 7: Madrid, Spain, girls and women aged 14–20: Bernis *et al.*, 1996

On the other hand, reproductive ageing in women is a unique process, in which the decrease in the capacity of adjustment to a system is connected to function loss. From an evolutionary perspective it is difficult to imagine that a process which is supposed to have originated as a specific adaptation does not have self-regulating mechanisms to allow for a normal transition to the loss of reproductive capacity.

The estimation of the average age of menopause is hindered by methodological difficulties. In addition, the range of variation in age at menopause is much wider in any population than the variation in age at menarche. For example, in our urban sample of women from Alcobendas, Madrid, the range of variation was 7 years for menarche and 27 for menopause (Bernis, 1995).

When comparing women who have developed their life cycles in very different ecosystems, it becomes apparent that ecological conditions associated with intense physical activity and low caloric intake reduce age at menopause, and vice versa (Beall, 1983; Bernis, 1973; Varea, 1990a, 1990b). Added to this, there is evidence that within a given population, women of high economic status have later menopause (Audit, 1964). For some populations, a secular delay in menopause has been demonstrated (Boldsen and Jeune, 1990).

The scarce data available for Morocco and Spain also point in this direction; the Moroccan women have an earlier menopause and among the Spanish women a certain variation in age at menopause exists and is associated with living standards

(Bernis, 1995). Moroccan women show later menarche and earlier menopause than the Spanish women, and within the Spanish groups for which we have information those that show later menarche also have earlier menopause. To a certain extent, this result is not surprising, as both menarche and menopause are ovarian function indicators. As such they must be shaped by the same range of ecological factors and show similar responses with regard to them.

Populations with higher caloric intake and lower energy expenditure levels, attributable to their low level of physical activity, have higher sex hormone levels than women with low caloric intake and high energy expenditure. Such conditions could account for earlier menarche and later menopause (Ellison and Lager, 1987; Gray, 1982; Rosetta, 1990), thus reducing the potential reproductive life span. Perhaps ovarian ageing will come to be considered as a biological marker of overall ageing, if the suggestion of Snowdon and colleagues (1989) is confirmed by further research.

The interaction of these biological patterns of maturation and ageing with different patterns of reproduction can give rise to different health problems in populations with late menarche and early motherhood, and late last maternity and early menopause. The evaluation of such aspects could enhance our understanding of steeply decreased fecundity and subsequent fertility, which has been described for women in both early post-menarcheal and late pre-menopausal phases. The demographic consequences of such variations would be open to research.

The socially accepted ages for beginning and ending reproduction may have very different consequences for women with different rates of sexual maturation and reproductive ageing (Frisancho *et al.*, 1984; Komlos, 1989; Ellison, 1991). In the Moroccan population of Amizmiz, women with late menarche and early marriage show a significantly lower prevalence of pregnancies during the first year of marriage, and have, besides, more abortions in their first pregnancy (Varea *et al.*, 1993). In a sample of rural Spanish women from Lugo, those who were adolescent mothers and late maturers produced higher frequencies than the other women of newborns with low birth weight and low cephalic circumference (Bernis and Varea, 1991).

One of the burning questions in Western societies at present concerns the delay in first maternity which is now approaching 30 years of age, and seems to be associated with increased difficulties in becoming pregnant and/or in complete gestations. The causes of the fertility decrease as menopause approaches have always given rise to discussion. A physiological explanation has been proposed, related to rapid reproductive ageing, of ovarian function itself and/or of the endometrial wall; this ageing determines conception difficulties (Treolar, 1974; James, 1979; Menken *et al.*, 1986; Frank *et al.*, 1991). Behavioural explanations also exist, related to a decrease in the frequency of sexual relations, as the age of the couple increases (Bongaarts and Potter, 1983). Voluntary termination, attributable to some fact of social significance, such as becoming a grandmother, has also been proposed (Caldwell and Caldwell, 1977; Naber, 1989).

The low success rate of artificial fertilisation in women over 30 is normally used

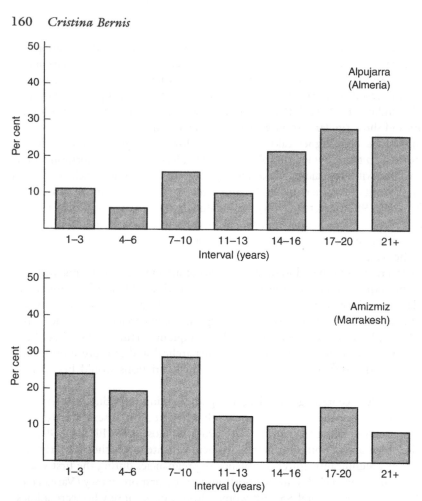

Figure 7.4 Distribution of the interval between the birth of a last child and onset of menopause among women in urban Spain and in rural Morocco

to give strength to the biological point of view. However, it is unlikely that the same factors influence the probability of a first pregnancy from 30 onwards and pregnancies with higher parities. Figure 7.4 shows the distribution of the intervals between the birth of a last child and age of menopause among rural Moroccan and urban Spanish women; more than 35 per cent of the Moroccan women had their last child no longer than 6 years before menopause. It is surprising that no more than 16 per cent of urban Spanish women fall within this range, in spite of their reduced family size.

In the 1970s, the percentage of Spanish women in their forties becoming mothers showed a considerable geographical variability, ranging from more than 10 per cent in Andalucia and Castille to less than 5 per cent in Cataluña. In 1989, the prevalence of mothers in their forties was reduced to 2.3 per cent in Madrid, and 2.7 per cent in Spain as a whole.

Among the rural Moroccan women who had completed their fertile life, 48 per cent of women became mothers between 35 and 44; and 15 per cent of them had a child at 45 years or later. We might ask whether conceptions before 30 years of age increase the capacity to conceive later in life, compared to the women who have not given birth by this age. If this is the case, we must enquire what explanation could be offered for such variations.

Behavioural determinants of fertile life use

There are two large groupings of behavioural determinants which operate during the reproductive phase. One group regulates the onset and end of the reproductive phase and final family size. The other determines the spacing patterns between successive children, regardless of the spacing methods used. Ultimately both sets of behavioural determinants depend on those value systems of each population in relation to the social role of women and their legal and personal situation.

Table 7.1 outlines the differences in reproductive patterns between populations where traditional norms characteristic of rural ecosystems (RE) predominate, such as Morocco, and populations in which modern norms characteristic of urban ecosystems (UE) predominate, such as Spain. The contrast is clear. In Morocco, modernisation of some urban nuclei and social sectors is being implemented very rapidly, while in Spain, although the country as a whole can be considered modern, it contains wide variations.

Figure 7.5, summaries studies by this author and others, demonstrating variabilities in biological and cultural determinants of the fertile life phase in different Moroccan and Spanish populations. A great difference is observed in the onset of the reproductive phase, and in the ages of first and last maternity. The big differences in family sizes and associated bio-social variables reflect the contrast that exists, in general, between Western populations and the economically developing world, with the exception that in Spain fertility control was introduced later than in other European countries, and in Morocco this is now occurring more rapidly than in other North African countries.

This different use of potential fertility related to reproductive patterns, such as age at marriage and family size, gives rise to differences in the biology of the women. For instance, Moroccan women experience during their fertile life phase only half the number of menstrual cycles than the Spanish women have: 214 and 411 respectively. This may have long term health effects at the end of reproductive life (Figure 7.6).

Fertility control is a priority policy today in Morocco, and the feasibility of the country's economic and social development plans greatly depends on its efficiency. However, fertility control implies not only an increase in contraceptive use, but also a change in traditional reproductive patterns. Such changes, unless properly monitored, can take their toll on the health and well-being of women. Could this further add to the high maternal mortality still apparent in Morocco?

Such concerns have invited our review and analyses of recent temporal and

Table 7.1 Main differences in reproductive patterns between rural and urban ecosystems

	Sexuality/Reproduction	*Women's status*	*Ideal family*	*Reproductive span*	
Rural ecosystems	Marriage, sexuality and reproduction go together	Universal marriage	Inside marriage Big (6 children)	Long: early age of first maternity, late age of last maternity	Maximum use of biological reproductive potential
Urban ecosystems	Separation of sexuality and reproduction, inside and outside marriage	Married or single	Inside/outside marriage Small (2 children)	Short: late age of first maternity, early age of last maternity	Restricted use of biological reproductive potential

Useful indicators to evaluate the transformation/modernisation in reproductive patterns

% single women sexually active	% of women married before 20	Mean family size	Mean age at marriage
% women using contraception	% of women single at 45	% of childless families	Mean age at first maternity
Age difference between mates	% single mothers	% of families with more than five children	% primiparity before 20
% of women married because of pregnancy			% primiparity after 30
			% last maternity before 30
			% last maternity after 40

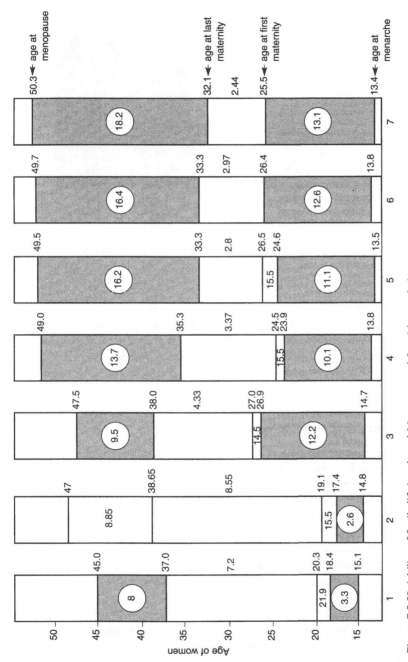

Figure 7.5 Variability of fertile life in selected Moroccan and Spanish populations
Sources

Col. 1: Rural Morocco: Verea, 1990
Col. 2: Marrakesh, urban Morocco: Crognier *et al.*, 1992
Col. 3: Leon, rural Spain: Bernis, 1974
Col. 4: Almeria, rural Spain: Verea and Bernis, 1991
Col. 5: Madrid, urban Spain: Bernis *et al.*, 1994
Cols. 6 and 7: Barcelona, urban Spain: Garcia-Moro and Hernandez, 1989

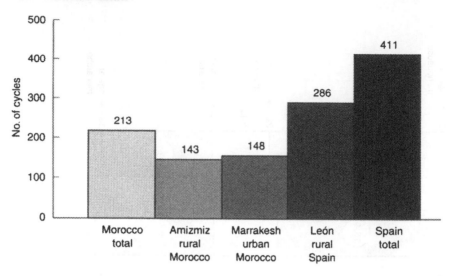

Figure 7.6 Mean number of menstrual cycles experienced by Moroccan and Spanish women

geographic variation in reproductive and fertility patterns in Spain. This could be used as a model for studying and preventing specific problems as they arise. The indicators used for this analysis draw on those that appear at the foot of Figure 7.5, to show the degree of modernisation or transformation of reproductive patterns (Bernis, 1991b). The data correspond to the 1985 fecundity survey in Spain (INE, 1986).

Fertility changes over time in Spain

Major declines in women's fertility rates in most European populations occurred during the nineteenth century. This has not been the case in Spain where such decline has occurred principally in the twentieth century (Coale and Watkins, 1986; Toharia, 1989). In Spain the most drastic reduction in the fertility rate occurred in the first half of this century, up to the time of the civil war. There was a second, less accentuated but more continuous reduction following the legalisation concerning contraceptives in 1978.

Marriage and fecundity within marriage continue to be the norm (Toharia, 1989), despite the increase in single motherhood, although this increase has never reached the levels found in other European countries such as Denmark (44 per cent) and the United Kingdom (25 per cent) (Eurostat, 1989).

Figure 7.7 shows the temporal variation in age at marriage, which increases slowly up to 1960 and declines gradually in both sexes. The difference in the age at marriage between men and women was narrowing by 1985. In the 1980s, a delay occurs in marriage age and in age at first maternity. In 1985, the age-specific fecundity rate is, for the first time, higher for women aged 30–34 than for women

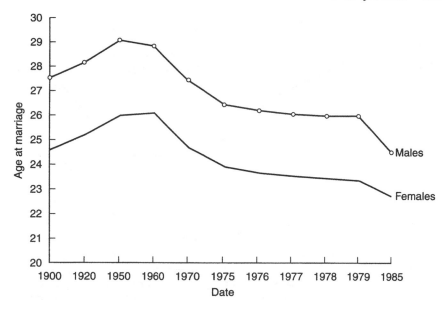

Figure 7.7 Variation in age at marriage in Spain, 1900–1985

in the 20–24 year age group. This illustrates how quickly proximate determinants of reproduction can change.

In Spain the impact on fertility of modern contraception and abortion is very difficult to evaluate before the late 1970s. This is because of the illegality of both prior to 1978. We do know that in 1973 only 120,000 women were legally prescribed the old contraceptive pill. These prescriptions were medically required by women because of serious health risks of potential pregnancy. Estimates of numbers of abortions prior to 1986 are also very problematic. In 1974 a magazine connected to the ecclesiastical hierarchy, estimated that about 300,000 women aborted illegally each year. These 'illegal' abortions were mostly carried out in the United Kingdom.

Socio-economic variation

Figure 7.8 shows the variability in some bio-social determinants of fecundity according to educational attainment in women. Women with a higher level of education married later and had children later. They also had their children over a shorter period of time.

We note here that when a comparison of family size with the level of education in the different autonomous regions is made, any differences disappear. That is to say, the variability in family size is greater by level of education of women than by the geographical area they belong to. Areas with higher fertility levels usually coincide with those where there is greater illiteracy and higher levels of primary

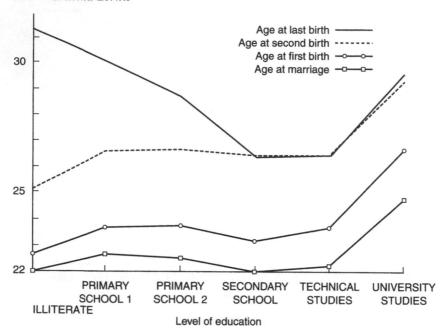

Figure 7.8 Variability in some bio-social determinants of fecundity according to the level of education of Spanish women

education only. These types of differential fertility and reproductive patterns, in relation to the level of women's education, are emerging in Morocco.

Methods of birth control: type, medical surveillance and failure

Data from the 1985 fecundity survey indicate that 49 per cent of the women between 18 and 49 years of age who were sexually active did not use any contraceptive method. Of those who did, 65 per cent did not monitor it. The need for surveillance undoubtedly depends on the contraceptive method being used. Data at a regional level indicate that the pill and withdrawal are used in equal proportions, 18 per cent. This indicates a change with regard to the 1977 survey when the prevalence was 17 per cent for pill use and 31 per cent for withdrawal.

Table 7.2 shows the type of contraceptive used in relation to birth regulation, previous to the last pregnancy, according to the educational level of women. Women with higher levels of education more frequently use contraceptive spacing methods in relation to their previous pregnancy. They more frequently use the pill, rhythm method and condoms, while those with lower levels of education more frequently use withdrawal. It is surprising, however, that, if some method of contraception is being used, its rate of failure varies little between levels of education.

Table 7.2 Type of contraceptive used in relation to the educational level of women

	Contraceptive used	Pill	Condom	Others	Contraceptive failed
Illiterate	37	4.6	0.8	31	18.3
No schooling	41	8.8	5.4	24	15.3
Primary	43	12	7.3	18	14.1
O-levels	49	18.5	9.3	14	13.0
A-levels	57	25.6	9.3	12	11.3
First degree	57	26	9.8	7	14.7
Postgraduate	66	28	11.5	12	13.7
Total	46	15	7.6	17	13.9

Table 7.3 Frequency of contraception use and its failure, by size of population

Population size	Contraception use	Contraception failure
Less than 10,000	72.3	32.4
10,001 – 50,000	62.7	35.9
50,001 – 500,000	67.4	32.0
Over 500,000	63.9	29.4
Total	69.9	32.4

The prevalence of contraceptive use in Morocco is now 35 per cent, and of these women, 95 per cent use the pill to regulate their fertility. This prevalence is even higher than that for the Netherlands. The level of medical surveillance is minimum, and those politically responsible for family planning are now trying to diversify the methods of birth control (Khachcha, 1994).

It is worthwhile to study the contraceptive methods used as well as their efficiency in use. The term 'effective method of contraception' is widely used to describe artificial modern methods, and 'inefficient method' to describe traditional contraceptive practices (Lighbourne *et al.*, 1982; National Research Council, 1989). We must be aware that the demographic transition of fertility rates in Europe was achieved using traditional contraceptive methods, long before the commercialisation of the 'modern' ones (Bernis, 1991b).

Table 7.3 shows the frequency of women who do not monitor their pregnancies as well as the frequency of contraception failure for married women, according to area populations. Those centres with fewer than 10,000 inhabitants show that significantly more women do not control their fertility. The efficiency of method used varies only slightly with the population size of communities.

In Spain, the frequency of pregnancies while using contraception is 11 per cent for married women. There are noteworthy regional and age differences. Contraceptive failure is usually higher among the youngest women, aged between 18 and 24. It is strongly associated with early marriage, frequently seen as a necessary solution to such social circumstances. Figure 7.9 shows the variability of the

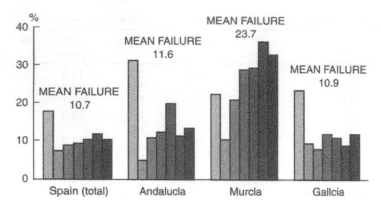

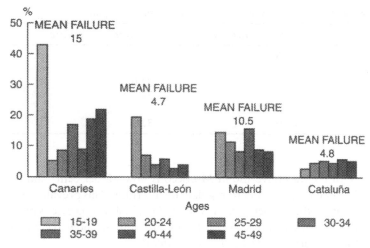

Figure 7.9 Age distribution of contraceptive failure in selected Spanish populations
Note
Age groups are 15–19, 20–24, 25–29, 30–34, 35–39, 40–44, 45+

pattern of contraceptive failure for some Spanish populations. Differences are astounding, not only in the total rate of failure, for instance 28.7 per cent in Murcia contrasting with 4.8 per cent in Catalonia, but also in age distribution. Most of the population show the highest failure rates among adolescents, as indicated by the bar graph. The opposite is true for some populations, and also for older women, those between 40 and 49 years, represented by the last two bars in the graph.

The government of Morocco has been promoting the use of contraceptive methods for years. The oral contraceptive pill has been the principal method employed. The prevalence of pill use is nearly double that of Spain, but the demographic impact is not yet spectacular.

Fertility control by the use of traditional methods has been accomplished in an effective manner in all the populations where it was socially acceptable. For example, Amizmiz women who marry young and who have a large family while still young, control their fertility in an effective way through traditional methods, as Varea (1993) showed.

Bley and Baudot (1986) in evaluating the efficiency of pill use in the Amizmiz community, found that the women with the largest family sizes were those using more frequently the oral contraceptive pill. Data from the World Fecundity Survey also point in this direction. In the initial transition phase, the pill and other modern methods are substituted for traditional ones in order to limit the increase of an already large family. Only in later phases of the demographic transition are these modern methods used to space and attain a previously planned, small family size.

In Morocco, and other Third World countries, use of the pill and other effective contraceptive methods is encouraged to control fertility. In Spain, as in the rest of Europe, the availability of the oral contraceptive pill has made possible the cultural separation of women's sexuality and fertility.

Specific health problems of women have been related to reproduction throughout biological history. For most of this time, childbirth and its consequences represented a high risk for women's lives and affected all social levels equally. Historical documents concerning the reproductive lives of the queens and princesses in Spain are shocking (Junceda, 1991). It is estimated that the maternal death rate of the reigning dynasties in Europe during the seventeenth and eighteenth centuries was around 2,000 per 100,000 (Tabutin, 1978).

In Morocco, the maternal death rate is around 300 per 100,000. This is much lower than the historical figure for European royalty, and also much lower than that of current sub-Saharan African and Asian populations, although it is much higher than the rate for Spain which is 15 per 100,000.

Different factors are associated with the risk of mortality during childbirth: the age of the mother, the level of hygiene, and primary health care support in its broadest sense.

It is usually argued that a reduction in fertility rate almost invariably reduces maternal mortality (National Research Council, 1989). However, if such a decline is not accompanied by drastic changes in the situation of poverty and lack of quality health facilities, by the implementation of women's enrolment for at least primary and secondary school education, and by the availability of birthing assistance and ante- and post-natal medical care, it is very unlikely that a reduction in maternal mortality will occur (Ahlberg, 1991; Winikoff, 1988).

In Morocco, although a great effort is being made to improve this situation, most women still give birth at home. There is a noteworthy contrast between the rural and urban environments.

Health consequences of change in reproductive patterns

Because the health problems that most affect women have been and still are related to reproduction or reproductive biology, it is not surprising that changes in patterns of sexuality and reproduction will affect women's health in different ways, thus changing the type and prevalence of health problems.

For women living in traditional rural ecosystems, the principal risk has been maternal mortality, while in modern urban settings, reproductive cancers and other diseases related to the reproductive system are the major causes of death and/or concern. The risk of certain reproductive cancers has a strong association with both late motherhood and repeated maternity. Further than this, such malignancies appear to be associated with changes in ecological conditions and lifestyles which include not only changes in reproductive schemes, but also changes in nutritional patterns and in women's social role. Social role change here could include incorporation in the labour force, the assumption of 'traditional male' behaviours, and stress patterns linked to modern urban life.

The very strong influence of commercial advertising on women's self-perception is giving rise to an increase in women dieting, an increase in moderate to heavy exercise by women, the use of aggressive contraceptive methods, and an increase in smoking, drinking, drug taking and recourse to plastic surgery. The internationally respected journal *Nature* once dedicated its front page to an article on beauty (Etcoff, 1994). This questioned beauty as an arbitrary cultural convention, and suggested that the general geometric features of a face give rise to a perception of beauty, may be universal, and may be governed by circuits shaped by natural selection in the human brain!

Generally speaking, changes in the environment concerning patterns of nutrition and infectious diseases have affected both sexes equally. However, changes in the socio-cultural environment, including technology, behaviour and social roles, have changed much more for women than for men. It is important to stress this point, because men are more sensitive than women to environmental change, although the intensity of the change for women has been greater. The bio-sociological and health consequences are far reaching.

Women are now putting many of their biological processes into the hands of the medical profession, for the control of dysfunctional menstrual cycles, fertility, the menopause, pregnancy, and childbirth. As evolutionary ecologists we may wonder how much intervention is wise. It is difficult to imagine that such an important biological function as reproduction does not have regulatory mechanisms available and capable of responding to wide ranging environmental changes.

Two explanations are possible: one that the pharmaceutical industry, and some medical professionals, are intervening too much in woman's biology, as a profitable business (Evans, 1979; Teo, 1990; Sloane, 1983; MacPherson, 1990), and normal biological processes are being inappropriately interpreted and treated as diseases. Although this may be true to a certain extent, we may wonder whether the other possible explanation is more likely. It may be that the degree of environmental change could have affected the reproductive biology of women so as to result in large mal-adaptations because the biology of reproduction was selected in biological and environmental circumstances which were completely different from those now prevalent in Western populations. Perhaps the high prevalence of dysfunctions related to menstrual cycles, pregnancies, the transition at menopause, and gynaecological cancers, that characterise Western women, are the signs of this mal-adaptation. If this is so, it is time to make a global analysis of the problem and to discuss possible solutions, such as monitoring the biological and behavioural proximates of fertility, and consequently perhaps diminishing the risk of suffering these dysfunctions and pathologies. Among these, age at menarche and age at first maternity rank high as relevant risk factors, and are probably the easiest to monitor, with the greatest potential benefits for women's health.

Age at menarche is a biological proxy, and has been shown to be a risk factor for adolescent pregnancies, menstrual dysfunctions (Bernis *et al.*, 1994b; Bernis, 1995), obesity and related health problems (La Velle, 1994) and breast cancer (Sherman *et al.*, 1981; Apter and Vihko, 1983; Micozzi 1985; Baanders and de Ward, 1992). As already discussed, late first maternity is a risk factor for difficulties in conceiving, for full term pregnancies (Velde, 1994) and also for breast cancer (Standford *et al.*, 1987; Lund, 1991).

Trying to delay age at menarche and to encourage early motherhood is a challenge we have to face, and is not an easy one. Monitoring menarche by nutritional and physical activity indices in childhood and adolescence could be easier than convincing women to have children, and to have them young, because political decisions are needed to implement social services, help for mothers and campaigns explaining the benefits for early motherhood (Velde, 1994). The question is open to debate, and in any case, we have to decide.

References

Anderch, B. and Milsom, I., 1982, An epidemiological study of young women with dysmenorrhea, *American Journal of Obstetrics and Gynaecology*, 655–660.

Audit, F., 1964, De quelques facteurs influençant la menopause, *Gynecologie Pratique*, 429–434.

Ahlberg, B.M., 1991, *Women, Sexuality and the Changing Social Order, The Impact of Government Policies on Reproductive Behaviour in Kenya*, Gordon and Breach.

Apter, D. and Vihko, R., 1983, Early menarche, a risk factor for breast cancer indicates early onset of ovulatory cycles, *Journal of Clinical Endocrinology and Metabolism*, 57: 82–88.

Baanders, A.N. and de Ward, F., 1992, Breast cancer in Europe: the importance of factors operating at an early age, *European Journal of Cancer Prevention*, 1: 285–291.

Beall, C., 1983, Ages at menopause and menarche in a high-altitude Himalayan population, *Annals of Human Biology*, 10: 365–370.

Bernis, C., 1973, Variaciones en la edad de menarquia y menopausia en una comarca natural (La Maragateria, León), *Trabajos de Antropologia*, 17: 7–19.

Bernis, C., 1980, Edades de menarquia y menopausia en la mujer española, *Actas del I Congreso Español de Antropologia*, Barcelona, 475–481.

Bernis, C., 1991a, Biological and behavioural aspects of human reproduction in two rural populations: Amizmiz (Marrakesh, Morocco) and la Maragateria (León, España), *Journal of Human Ecology*, 1, 1: 63–75.

Bernis, C., 1991b, Global Changes and Their Implications for Women's Health, *Journal of Human Ecology*, 2, 1/2: 171–195.

Bernis, C. and Varea C., 1991, Maternal factors and developmental variables of newborns, Lugo, España, *VI International Congress of Auxology*, Madrid.

172   *Cristina Bernis*

Bernis, C., Varea, C. and Robles, F., 1994, Edad de menarquia y ciclos menstruales en adolescentes, in: *Biologia de poblaciones humanas: problemas metodologicos e interpretación ecológica*, 803–815. Ediciones Universidad Autonoma, Madrid.

Bernis, C., 1994, Changement dans les regulateurs culturels et sociaux de la vie feconde des femmes espagnoles et marocaines: implications biologiques et consequences pour la santé. (Unpublished paper.)

Bernis, C., 1995, Procesos biologicos y estilos de vida en mujeres de Alcobendas (Unpublished paper.) Universidad Autonoma, Madrid.

Bley, D. and Baudot, P., 1986, Some recent trends in infant mortality in the province of Marrakesh, Morocco: a demographic transition in process, *Social Biology*, 33: 322–325.

Boldsen, J. and Jeune, B., 1990, Distribution of age at menopause in two Danish samples, *Human Biology*, 62(2): 291–300.

Bongaarts, J. and Potter, R., 1983, *Fertility, Biology and Behaviour, An analysis of the proximate determinants*, Academic Press.

Caldwell, J. and Caldwell, P., 1977, The role of marital sexual abstinence in determining fertility: a study of Yoruba in Nigeria, *Population Studies*, 31: 193–199.

Coale, A.J. and Watkins, S. (ed.), 1986, *The Decline of Fertility in Europe*, Princeton University Press.

Crognier, E., Bernis, C., Elizondo, S. and Varea, C., 1992, Reproductive patterns as environmental markers in rural Morocco, *Coll. Anthrop*, 16, 1: 89–97.

Direction de la Statistique, 1984, *Caracteristiques Socioeconomiques de la Population*, Rabat.

Donalson, L., 1991, *Fertility Transition. The Social Dynamics of Population Change*, Blackwell.

Etcoff, N., 1994, Beauty and the beholder, *Nature*, Vol. 368, March: 186–187.

Ellison, P. and Lager, C., 1987, Low profiles of salivary progesterone among college undergraduate women, *Journal of Adolescent Health Care*, 8: 204–208.

Ellison P., 1991, Reproductive ecology and human fertility, in: Mascie-Taylor, N. and Lasker, G. (eds), *Applications of Biological Anthropology to Human Affairs*, Cambridge University Press, 14–54.

Eurostat, 1989, *Boletin de Estadística Europeo*.

Evans, T., 1979, Ten years' review of hysterectomies: trends, indicators and risks, *American Journal of Obstetrics and Gynaecology*, 91: 73–77.

Frank, O., Bianchi, G. and Campana, A., 1991, The end of fertility: age, fecundity and fecundability in women, *Journal of Biosocial Science*, 26(3): 349–368.

Frisancho, A., Matos, J. and Flegel, P., 1984, Role of gynaecological age and growth and maturity status in foetal maturation and prenatal growth of infants born to young still growing adolescent mothers, *Human Biology*, 6: 583–594.

Garcia-Moro, C. and Hernandez, M., 1990, Changes in age at menarche in Spain (1909–1965), *International Journal of Anthropology*, 5, 2: 117–124.

Gray, B., 1982, Enga birth maturation and survival: Physiological characteristics of the life cycle in the New Guinea Highlands, in: MacCormack, C.P. (ed.), *Ethnography of fertility and birth*, 75–114.

Henry, L., 1979, Concepts actuelles et resultats empiriques sur la fecondité naturelle, in: Leridon, H. and Menken, J. (eds), *Fecondité Naturelle*, 17–28.

Holloway, M., 1994, Trends in women's health, *Scientific American*, August, 67–73.

Instituto Nacional de Estadística (INE), 1986, *Encuesta de Fecundidad, 1985*, Instituto Nacional de Estadística, Madrid.

Jain, A. and Bruce, J., 1993, *Implications of reproductive health for objectives and efficacy of family planning programs*, Population Council, Working papers, No. 8, New York.

James, W., 1979, The causes of the decline of fecundability with age, *Social Biology*, 26: 330–334.

Junceda, E., 1991, *Ginecologia y vida íntima de las reinas de España*, Vol. I (ed), in series Temas de Hoy.

Khachcha, M., 1994, Evolution du concept de la planification familiale au Maroc et perspectives d'avenir, in: Crugnier, E., Baali, A. and Boetoh, G. (eds), *Conception, Naissance et Petite Enfance au Magreb: Approches Anthropologiques* (in press).

Komlos, J., 1989, The age at menarche and the age at first birth in an undernourished population, *Annals of Human Biology*, 16: 463–466.

Lalonde, M., 1974, *A New Perspective on the Health of Canadians*, Ottawa.

La Velle, M., 1994, Are there long Term Consequences of Secular Trend in Early Menarche? *Collegium Antropologicum*, 18(1): 53–63.

Lighbourne, R., Sing, S. and Green, C., 1982, The world fertility survey: charting global childbearing, *Populations Reference Bureau*, 37, 1.

Lund, E., 1991, Breast cancer mortality and the change in fertility risk factors at menopause: a prospective study of 800,000 married Norwegian women, *Epidemiology*, 2, 4: 285–288.

MacPherson, K., 1990, Nurse-researchers respond to the medicalisation of menopause, in: Flint, M., Kronenberg, F. and Utian, W. (eds), *Multidisciplinary Perspectives on Menopause*, 180–184.

Menken, J. and Trussell, J., Larsen, U., 1986, Age and infertility, *Science*, 233: 1389–1392.

Micozzi, M., 1985, Nutrition, Body Size and Breast Cancer, *Year Book of Physical Anthropology*, 28: 175–206.

Naber, N., 1989, *Etude du comportement fécond d'une population féminine de la haute vallée d'Azgour* (Amizmiz, Marrakesh), doctoral thesis, University of Caddy Ayyad, Marrakesh.

National Research Council, 1989, *Contraception and reproduction: Health consequences for women and children in the developing world*, National Academy Press, Washington DC.

Omran, A.R., 1971, The Epidemiologic Transition: A Theory of the Epidemiology of Population Change, *Millbank Memorial Foundation Quarterly*, 49.

Prado, C., 1984, Secular change in menarche in women in Madrid, *Annals of Human Biology*, 11(2): 165–166.

Prado, C., 1992, Variation seculaire de la menarche chez une population du nord du Maroc: aspects socioeconomiques et nutritionelles, *Revista di Antropologia*, LXX: 139–145.

Rogers, R.G. and Hackenberg, R., 1987, Extending epidemiologic transition theory: a new stage, *Social Biology*, 34(3–4): 324–343.

Rosetta, L., 1990, Biological aspects of fertility among third world populations, in: Landers, J. and Reynolds, V. (eds), *Fertility and Resources*, Cambridge University Press: 18–34.

Sherman, B., Wallace, R., Bean, J. and Schlabaugh, L., 1981, Relationship of body weight to menarcheal age and menopausal age: implications for breast cancer risk, *Journal of Clinical Endocrinology and Metabolism*, 52(3): 488–493.

Singh, S. and Ferry, B., 1984, Biological and traditional factors that influence fertility: results from the world fertility surveys, *Comparative Studies*, 40.

Sloane, E., 1983, *Biology of Women*, Delmar, NY.

Smith, R., 1983, Evolutionary ecology and the analysis of human social behaviour, in: Dyson-Hudson, R. and Little, M. (eds), *Rethinking Human Adaptation. Biological and cultural models*, Westview Press, 23–40.

Snowdon, D.A., Kane, R.L. and Beeson, W.L., 1989, Is early menopause a biological marker of health and ageing? *American Journal of Public Health*, 79(6): 709–714.

Standford, J., Hartage, P., Brinton, L., Hoover, R. and Brookmeyer, R., 1987, Factors influencing the age at natural menopause, *Journal of Chronic Diseases*, 40(11): 995–1002.

Tabutin, D., 1978, La surmortalité feminine en Europe avant 1940, *Population*, 1: 121–148.

Teo, P., 1990, Hysterectomy: a change of trend or a change of heart?, in: Roberts, (ed.), *Women's Health Counts*, Routledge, London.

Treolar, A., 1974, Menarche, menopause and intervening fecundability, *Human Biology*, 46(1): 89–107.

Toharia, J.J., 1989, *La mitad de la explosión*, La población Española en perpectiva comparada, Fundación Banco Exterior, Madrid.

United Nations, 1991, *Human Development Report 1991*, Oxford University Press.

Varea, C., 1990a, La etapa final del periodo reproductor femenino: biologia y sociedad en una población del Marruecos rural (Amizmiz, Marrakesh), in: Bernis, C., Demonte, V. and Garrido y Calbet, T. (eds), *Los estudios de la mujer, de la investigación a la docencia*, Universidad Autonoma, Madrid.

Varea, C., 1990b, Patrones reproductores y fertilidad en una población tradicional de Marruecos (Amizmiz, Marrakesh), doctoral thesis, Universidad Autonoma Madrid.

Varea, C., 1993, Marriage, age at last birth and fertility in a traditional Moroccan population, *Journal of Biosocial Science*, 25:1–2.

Varea, C., Bernis, C. and Elizondo, S., 1993, Physiological maturation, reproductive patterns, and female fecundability in a traditional Moroccan population (Amizmiz, Marrakesh), *American Journal of Human Biology*, 5: 297–304.

Velde, E.R., 1994, Pregnancy in the 21st century: consistently later, consistently more artificial. Lecture delivered in Nov. 1991, Universidad Autonoma, Madrid.

Vihko, R. and Apter, D., 1984, Endocrine characteristics of adolescent menstrual cycles: impact of early menarche, *Steroid Biochemistry*, 20, 231–236.

Winikoff, B., 1988, Women's health as an alternative perspective for choosing interventions, *Studies in Family Planning*, 9(4): 197–214.

WHO, 1985, *Targets for Health for All*, WHO Regional Office for Europe, WHO European Series for HFA, No. 1.

WHO, 1988, *Priority Research for Health for All*, WHO Regional Office for Europe, WHO European Series for HFA, No. 3.

8 Health of children

Causal pathways from macro to micro environment

Nick Spencer

Abstract

Child health is determined by a complex interplay of factors. Conflicting explanations for differences in child health outcomes between and within child populations have been advanced: one large school of thought has focused on parenting and variables within the child's micro environment; the opposing school has focused on the macro environment in which child populations live with particular emphasis on the socioeconomic determinants. This chapter explores the relationship between macro and micro environmental influences on child health using data from a range of studies and the causal pathways by which socio-economic and socio-cultural variables in the macro environment may influence the child's micro environment. The concept of proximal and distal causes is used to consider causal models. The chapter concludes by proposing a framework for the use of causal pathways in the study of child health outcomes.

Introduction

The main determinants of child health have been the subject of a long and intense debate (Spencer, 1996). The chief protagonists advance apparently conflicting explanations for child health outcomes: one school of thought focuses on the immediate environment of the child influenced mainly by parental health-related and culturally determined behaviours; the other school is concerned with structural and material influences which are mainly centred outside the child's immediate environment and are without parental control. In ecological terms, the former is concerned with the micro environment and the latter with the macro environment.

These explanations are frequently seen as mutually exclusive and data analyses, following the 'germ theory' on which much of modern medicine is based, are directed to finding the single most important determinant using statistical techniques which weight the relative strength of co-variables (Logan and Spencer, 1996). This pursuit of single causative agents or correlates has not only generated many misleading conclusions but is fundamentally incapable of accounting for the complexity of the processes by which child health is determined (Rutter, 1988).

Causal agents such as bacteria or smoking which exert a direct pathological effect on the child are themselves influenced by other variables – what Rose (1992) terms 'causes of causes'. These can be seen as 'proximal' and 'distal' causes (ibid.), the proximal operating in the micro environment and the distal in the macro environment. Causal pathways linking these variables are likely to provide models for explaining the determination of child health capable of accounting for the complexity of the relationship in terms both of proximity and temporal and inter-generational effects, the importance of which has been increasingly recognised in recent years (Barker, 1992).

This chapter explores the causal pathways which link micro and macro environmental influences on child health, concluding with a theoretical framework for the construction of such pathways and a discussion of the research and environmental policy implications of the focus on pathways rather than single causal agents. Before specifically considering examples from the literature of causal pathways in child health, the chapter briefly reviews micro and macro environmental explanations for child health outcomes and considers those variables which might mediate between the micro and the macro environment.

Brief overview of micro environmental explanations

Micro environmental explanations of child health outcomes centre on the parents, particularly the mother, and the physical and emotional environment in which they nurture their children. Variables related to this immediate environment have been widely studied; the level of hygiene and cleanliness of the home (Spence *et al.*, 1954; Moy *et al.*, 1991; Paton and Findlay, 1926); maternal 'inefficiency/incompetence' (Paton and Findlay, 1926; Burns, 1947; Douglas, 1951); parental (usually maternal) stimulation of the child and parenting styles (Fergusson *et al.*, 1984); the mother's education and her level of 'ignorance' or enlightenment (Caldwell and McDonald, 1982); the child's psychological environment determined by the mother's own mental health (Keitner and Miller, 1990) and the parental experience of being parented (Browne, 1988); nutrition of the child (Aukett and Wharton, 1995); health related behaviours such as smoking (Poswillo and Alberman, 1992); culturally determined child care practices (Bailey *et al.*, 1990).

The variable representing hygiene and cleanliness of the family home has been correlated with higher levels of childhood mortality and morbidity (Spence *et al.*, 1954; Moy *et al.*, 1991), as has 'maternal inefficiency/incompetence' (Paton and Findlay, 1926; Douglas, 1951). Parental stimulation, parenting styles and parental depression are associated with developmental delay and child abuse and neglect (Lyons-Ruth *et al.*, 1990). Maternal education is linked with infant mortality and various measures of health outcome (Caldwell and McDonald, 1982), and maternal smoking has been linked with a range of child health outcomes from perinatal mortality (Butler *et al.*, 1972), Sudden Infant Death Syndrome (Mitchell *et al.*, 1992), respiratory illness (Evans and Golding, 1992), hospital admission (Harlap and Davies, 1974) and developmental delay (Fogelman, 1980).

These variables, which have been studied in widely differing historical and global settings related to a range of child health outcomes, have in common their proximity to the child and their close relationship to the individual attributes of the parents, in particular the mother. Concepts such as 'maternal inefficiency/incompetence' are generally discredited and no longer in use; however, these micro level variables tend to be studied in isolation from the socioeconomic context and are broadly viewed as the individual responsibility of the parent or the result of wider cultural factors independent of economic and other environmental influences (Finerman, 1994).

The consequence of this individual, micro level focus is that preventive solutions are couched in terms of individual level interventions usually in the form of education. Health promotion related to maternal smoking in pregnancy is a good example of this approach; various health education campaigns have stressed the individual choice facing pregnant women even though the efficacy of such campaigns has been questioned (Madeley *et al.*, 1989) and others have demonstrated the need to place smoking cessation within a social context if they are to address the realities of the pressures on mothers to smoke (Blackburn and Graham, 1993). In less developed countries, the solutions are seen as education (Caldwell and McDonald, 1984) as well as attempts to counter traditional child care practices (Finerman, 1994).

The need for education, both in terms of overcoming ignorance and in responsible parenting, was among the recommendations of the Inter-departmental Committee set up by the UK government in 1904 in response to the poor physical state of young male recruits at the time of the Boer War. The Committee's conclusions were influenced by professional opinions expressed by those such as a voluntary health visitor who stated:

> The girls . . . have no sort of sense of duty; not the slightest. It is only amusement and pleasure with them. The last thing they think of is duty, and therefore, they do not trouble to cook or get up in the morning and the children go to school without breakfast, because the woman is too idle to get up . . . she is utterly indifferent.
>
> (Smith and Nicholson, 1992)

Some years later, Golding and Butler (1984) reporting on their analysis of the results of the second of the UK national cohort studies, reach similar conclusions in relation to the child health inequalities noted in the study although they directly implicate smoking and alcohol:

> There is a feeling that because there are strong trends in perinatal and neonatal death with social class, political action to equalise the wealth or the housing of the lower classes would automatically result in an equalising of the death rate. From the available evidence this is unlikely to happen. As we have shown, a major determinant of the differences in death rates is the maternal smoking history – give the family more money and it is possible that their consumption of tobacco (and alcohol) would increase.

These quotations illustrate the assumptions underlying the focus on the attributes and behaviour of individual mothers. The possibility of poor socioeconomic circumstances influencing the ability to parent and provide adequate nurture appears not to have occurred to the voluntary health visitor and is dismissed out of hand by Golding and Butler.

In summary, micro level explanations tend to decontextualise behaviours and parental attributes and behaviours, seeing them as independent of social and economic pressures. As this chapter aims to illustrate, the effects of micro level variables on child health can only be understood within a wider social, economic and cultural context.

Brief overview of macro explanations

Macro environmental explanations shift the focus from the individual to the broader societal level. Child health is seen as being determined primarily by social, political and economic forces outside the control of the individual and by the established structures of society which favour privileged minorities. These explanations carry the implicit message that child health is most likely to improve in response to social and economic changes which minimise poverty and favour the majority in countries over the narrow interests of privileged minorities.

National resources and the extent to which societies invest in education of women (UNHDP, 1994) as well as income distribution (UNHDP, 1994) and levels of relative and absolute poverty (Wilkinson, 1994; Grant, 1994) within countries are seen as influencing the health of children within that society. War and international economic and social policies which distort local economies and determine how resources are spent (SCF, 1995; Costello *et al.*, 1994) are thought to undermine national attempts to improve the health status of children. Access to safe water (Ebrahim, 1985), housing and adequate shelter (Ineichen, 1993), an unpolluted environment and access to transport and medical care all constitute important child health promoting factors which are beyond the control of the individual and are dependent on international and state resources and policies.

Factors in the macro environment advanced as health determinants may appear most applicable to less developed countries. However, it is argued not only that changes in these same structural and societal factors explain the dramatic improvements in health status in developed countries over the last hundred years (McKeown, 1976; Szreter, 1988), but that persisting structural inequalities account for much ill health among adults and children (Townsend *et al.*, 1992; Smith *et al.*, 1994; Jolly, 1990). The continuity of evidence for the effects of macro environmental factors on child health across countries and historical periods has been presented as confirmation of their power as explanatory variables in relation to child health status (Spencer, 1996).

These macro environmental factors have been linked with the same child health outcomes as micro environmental variables: poverty and low income are correlated with higher levels of perinatal and infant mortality (Kumar, 1993), low birth weight (Reading *et al.*, 1990), respiratory infections and hospital admissions

(Spencer *et al.*, 1993; Spencer *et al.*, 1996) and developmental delay (Wedge and Prosser, 1973; Sampson and Laub, 1994; Garrett *et al.*, 1994); damp housing is associated with increased respiratory symptoms (Martin *et al.*, 1987); lack of safe water is correlated with diarrhoea in less developed countries (Ebrahim, 1985).

Macro explanations have been criticised on the grounds that they are unable to explain why some children living in similar economic circumstances have different outcomes, thereby suggesting that other variables, at the micro level, must play a role. Some studies of less developed countries have concluded that socioeconomic factors have little influence on various child health outcomes such as infant mortality (Bailey *et al.*, 1990) and frequency of diarrhoeal illness (Moy *et al.*, 1991). The setting of these studies in universally poor villages may limit their capacity to examine socioeconomic differences in child health outcomes across whole populations (Millard, 1985); however, their findings do suggest that some children in poor families are protected in some way by micro environmental factors.

In summary, macro environmental explanations seem to account for some differences in child health status between social groups and between countries, and improvements in health in developed countries which have accompanied better nutrition and improved living standards owe more to macro than to micro environmental changes. However, there remain some unexplained differences in outcome between children experiencing similar macro environments. These may be partly explained by differences in the micro environment.

Mediators linking the macro to the micro

Micro and macro environmental factors both influence child health and have been correlated with the same child health outcomes. It may be that micro and macro environmental factors exert their effects independently, but the evidence suggests that they are closely linked by causal chains made up of a range of mediating variables operating at different levels. Although the micro environment is able to exert some influence on the macro, the general direction of the causal chains is likely to be from macro to micro. As an illustration, consider the disposal of faecal waste in villages without latrines. Defecation by individuals in the fields and in communal areas has a direct effect on the environment of others becoming part of wider or macro environmental health hazards. However, the absence of latrines and adequate sanitation in such villages is a function of the lack or maldistribution of resources in the societies of which the villages are a part and a maldistribution of resources between countries and continents.

There is a multiplicity of mediators between the macro and micro environments. In order to explore some of the most important mediators in the limited space available, I will consider those which are known to link the most frequently identified micro environmental factors associated with child health with the macro environment. In the developed world, maternal smoking and parenting skills, including stimulation and supervision, have been shown to influence child health outcomes at the micro level; hygiene has been the main micro environmental focus in less developed countries.

At first sight, smoking would appear to be simply a matter of personal choice and much health promotional material has approached it in this way, stressing the health hazards and urging individual mothers not to smoke especially during pregnancy (Graham, 1989; Golding *et al.*, 1992). In fact, smoking in pregnancy, and among mothers generally, has in the past 30 years become closely linked to material deprivation and disadvantage (Graham, 1993a; Chollet-Traquet, 1992). There is powerful evidence demonstrating the association between caring in adverse material and social circumstances and maternal smoking (Graham, 1993b). What are the factors which appear to mediate this link between the macro socioeconomic influences and smoking? The UK data show that mothers living on state benefits experience greater levels of depression and feelings of powerlessness in the face of the pressures of material hardship (Graham, 1993b). These women are also likely to be coping with the pressures of poor and inadequate housing (Martin *et al.*, 1987), dangerous physical environments in the home and the surrounding streets (Roberts *et al.*, 1995) and family debt (National Children's Homes, 1992).

As discussed above, dysfunctional parenting, 'maternal inefficiency/incompetence' and the level of parental stimulation and supervision of their child are micro level factors correlated with adverse physical and mental health outcomes in childhood. All these factors are in turn related to the socioeconomic status (SES) of the family and have been shown to change as family SES changes (Garrett *et al.*, 1994; Sampson and Laub, 1994). As with smoking, debt and coping with inadequate resources and maternal depression act as mediators in the chain from macro factors to parenting problems. Parenting for those in chronic poverty and disadvantage (Danziger and Stern, 1990) is most challenging and most likely to be dysfunctional especially in societies in which the poor are excluded from the experience of the majority of citizens (Wilkinson, 1994).

Maternal education level has been linked to parenting styles (Sampson and Laub, 1994) and many other child health outcomes. Education is often treated as an independent variable in socio-medical research; however, maternal education is itself partly determined by macro economic and societal factors (Palloni, 1981) and acts as a powerful mediator between macro and micro environmental factors. This is well demonstrated in less developed countries where differences in maternal education are powerfully linked with child health outcomes (Caldwell, 1979). Poverty and low SES determined at the macro level tend to exclude the poor from educational opportunities which, in turn, condemns them to lower paid work and greater chances of unemployment.

Poor hygiene in the home is linked with increased infections in childhood, particularly diarrhoea in the first few years of life (Bailey *et al.*, 1990; Moy *et al.*, 1991). Hygiene is linked with macro economic and environmental factors through the availability of clean water and sanitation and maternal education.

In the examples considered above, macro environmental influences on health are mediated through a series of other variables to contribute to the micro environmental factors which directly influence the child in the home environment. No attempt has been made to compile an exhaustive list of mediators between macro and micro environment but the examples serve to illustrate the working of the causal chains.

Temporal, inter-generational and cumulative effects

In considering causal processes linking macro and micro environmental factors, temporal and inter-generational effects need to be taken into account. Changes in the macro environment may take some time to exert their effect at the micro level as families and children are shielded from the effects by mediating and modifying variables. Equally, mediating factors may change but the effects of change may not be manifest for some time.

Nutritional changes experienced by the Dutch population during the 1944–45 Dutch Hunger Winter were abrupt and drastic (Hart, 1993) with an immediate effect on the micro environment. However, there is evidence that, in some parts of the Netherlands, the effects at the micro level were modified by inter-generational influences in that the foetuses of mothers in the areas most severely affected by the food shortage were partially protected by the superior nutritional status of children in these parts of the Netherlands which had been more prosperous for generations (Hart, 1993).

The effects of the economic depression of the early 1980s in the UK and other industrial countries are likely to have been more insidious. Despite a rapid rise in unemployment and an increase in families living on low income (Townsend *et al.*, 1992), evidence for the effects on poor communities and an increase in health inequality started to emerge only in the early 1990s (Phillimore *et al.*, 1994; McLoone and Boddy, 1994).

Macro environmental influences on child health can be exerted through inter-generational effects. Baird (1974) has demonstrated the relationship between the nutritional status of the female child usually manifest in her growth during childhood and the outcome of her pregnancy when she becomes a mother; the same relationship is responsible for the so-called cycle of deprivation in less developed countries from chronically undernourished mother to low birth weight and stunted child (Sanders, 1985). Adult health outcomes seem to be correlated with events during the foetal and infant life of the individual (Barker, 1992). There is also evidence that micro environmental factors can exert effects across generations; this is particularly evident in the inter-generational effects of child abuse and neglect (Browne, 1988).

Factors in the micro and macro environment exert their influence not only over time and across generations but also cumulatively. The presence of one adverse factor is frequently correlated with others which together exert a cumulative influence which may be greater than the sum of their individual effects. For example, UK children from families with multiple indicators of disadvantage studied at the age of 11 years as part of the 1958 National Cohort study (Davie *et al.*, 1972) were shown to be 13 centimetres shorter than children from families with none of these disadvantages, whereas the effects of single markers of disadvantage on height attained were much less marked.

The temporal, inter-generational and cumulative effect of micro and macro level factors is well-illustrated by the following quotation from Smith and colleagues (1994):

> A woman in a low-income household is more likely to be poorly nourished during pregnancy and to produce a low birthweight or premature baby. A child growing up in a low-income household is more likely to be disadvantaged in terms of diet, crowding, safe areas in which to play and opportunities for educational achievement. An adolescent from a low-income household is more likely to leave education at the minimum school-leaving age, with few qualifications and to experience unemployment before entering a low-paid, insecure and hazardous occupation, with no occupational pensions scheme. An adult working in this sector of the labour market is more likely to experience periods of unemployment, to raise a family in financially difficult circumstances and to retire early because their prematurely expended health can no longer cope with the physical demands of work. A retired person who does not have an occupational pension is more likely to experience financial deprivation in the years leading up to their death.
>
> (p. 140)

Causal pathways from macro to micro environment

As argued above, the concept of single causal agents, either in the micro or macro environment, is inadequate to the task of explaining the complexity of factors determining child health. Models, which attempt to account for the complexity and the inter-relationship of the independent variables and are able to incorporate temporal, inter-generational and cumulative effects, provide a more satisfactory framework for understanding child health determinants. Single causal agents have been investigated widely; by contrast, causal modelling in relation to child health is relatively under-researched. The examples used in this chapter are some of the exceptions; they are taken from less developed and developed countries and address a variety of outcomes from infant mortality to adolescent offending. Their unifying theme, for the purposes of this chapter, is their attempt to link the macro to the micro environment.

Millard (1994) proposes a model of the causes of high levels of child mortality in less developed countries (Figure 8.1). She utilises three tiers: the ultimate tier which includes the effects of international and national economic factors and natural ecology on household food security and settlement patterns; the intermediate tier, which includes micro environmental factors such as the distribution of food within the family and the diets of the children; and the proximate tier which is the cycle of malnutrition, diarrhoea and lower respiratory infection responsible for so much child mortality in less developed countries and in developed countries in the past (Sanders, 1985).

This is a theoretical model which is not based on specific study results and Millard makes no attempt to weight the variables or measure their relative

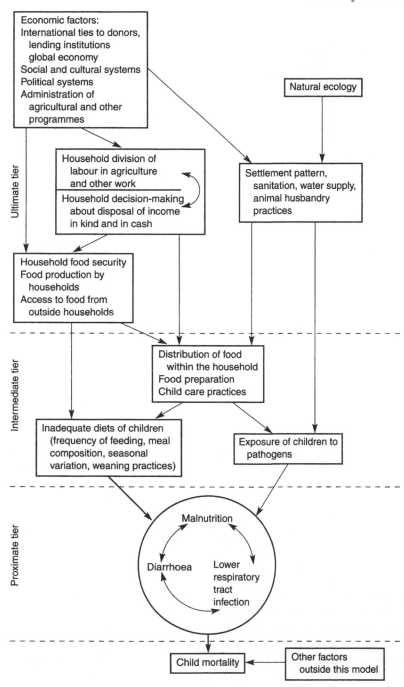

Figure 8.1 Three-tier causal model of child mortality
Source
Millard, 1994

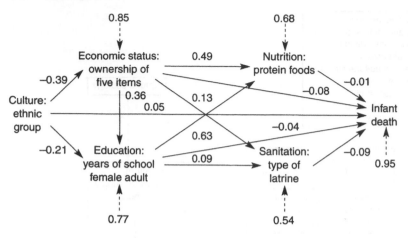

Figure 8.2 Path model showing cultural, socioeconomic and medical predictors of infant
death in 480 Sri Lankan households
Source
Waxler *et al.*, 1985

explanatory strength. Although her model does not make explicit reference to temporal, inter-generational and cumulative effects, the structure is flexible and adaptable enough to accommodate change over time and cumulative effects.

By contrast, Waxler and colleagues (1985) propose a path model based on a single study in Sri Lanka to explain some of the variance in infant mortality between households (Figure 8.2). This study found that various micro and macro environmental factors were associated with infant mortality but were also inextricably bound together. The resulting path model shows how these inter-relationships contribute to infant death. The authors acknowledge that the factors they have explored explain only 5 per cent of the variance. This is due both to their use of individual rather than aggregated data and to their choice of variables. Different levels of influence are not specifically acknowledged in this path model and no attempt is made to explore temporal and inter-generational effects. Despite the limitations of this study and the resulting model, the authors demonstrate the importance of the construction of causal pathways in trying to understand the relationship between macro and micro environmental factors and in unravelling the effects of co-variables.

Power and colleagues (1991) developed a theoretical model against which they tested their data on the inter- and intra-generational relationships between health and social circumstances (Figure 8.3). This model explicitly acknowledges temporal and inter-generational effects and suggests some cumulative influences. Proximal and distal levels at which the various variables operate are not specifically stated; however, the socioeconomic circumstances group of variables include both neighbourhood and 'period', and both of these imply a macro environmental influence in part dependent on the contemporary economic climate.

Frisch and colleagues (1992) studied infant survival in a neonatal intensive

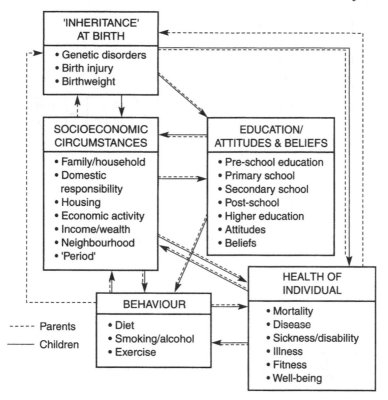

Figure 8.3 Inter- and intra-generational relationships between health and social
circumstances

Source
Power *et al.*, 1991

care unit. The authors use a path analysis method based on three criteria for inferring a causal relationship between variables and an outcome: the cause precedes the effect in time; the two variables are empirically correlated with each other; the correlation of the variables cannot be explained by a third variable that influences both of them. They construct a hypothetical model (Figure 8.4) using these three criteria and, based on this hypothetical model, use their data to construct a final path analysis model (Figure 8.5). These models demonstrate temporal and inter-generational effects on infant survival and the influence of macro environmental factors on the micro environment of the newborn infant.

Adolescent offending, in common with infant death, has been attributed to a variety of causes in both the micro and macro environment, from lack of parental supervision (Patterson and Stouthamer-Loeber, 1984) to unemployment (Farrington *et al.*, 1986). Based on the follow-up studies of the Newcastle (UK) Thousand Families Study (Miller *et al.*, 1960), Kolvin and colleagues (1988) explore the relationship of micro environmental factors such as mothering and

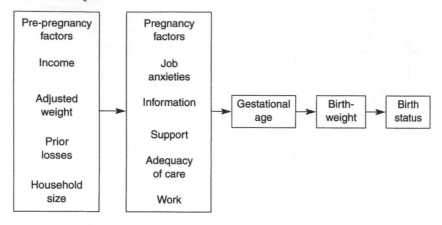

Figure 8.4 Hypothetical path to survival in a neonatal intensive care unit
Source
Frisch *et al.*, 1992

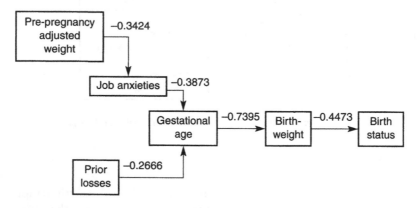

Figure 8.5 Final path analysis model
Source
Frisch *et al.*, 1992

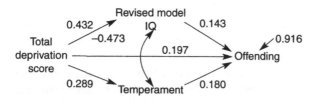

Figure 8.6 Model of relationship between deprivation, IQ, temperament and offending
        in adolescence
Source
Kolvin *et al.*, 1988

macro factors such as adverse social circumstances on adolescent offending using path analysis methods. They conclude that mothering and adverse social circumstances both have independent effects on offending and that marital disturbance, though a risk factor in its own right, is not so in the presence of severe multiple deprivation, and the effects of marital disturbance are mediated through a consequent poor quality of mothering and care and through impoverished social circumstances. The model (Figure 8.6) shows that the effects of deprivation on offending are mediated through prior effects on IQ and temperament.

These various examples illustrate attempts to use causal models to explore the complexity of relationships between different variables in the determination of child health. Though failing to explain fully the determinants of the aspects of child health under investigation, they move the explanatory debate beyond the confines of the search for the single causal agent and demonstrate the intimate and important links between macro and micro environmental factors in influencing child health outcomes.

A theoretical framework for the analysis of causal relationships in child health

The foregoing discussion demonstrates the problems and possibilities of constructing causal pathways in the study of child health determinants. Neither micro nor macro environmental explanations alone are sufficient to account for the complexity of child health determination. A better understanding to guide prevention and health promotion requires a theoretical framework within which causal relationships can be analysed. Such a theoretical framework is proposed below (Figure 8.7). The underlying principles of the framework are as follows:

- it must be able to incorporate the effects of a range of co-variables operating both in the macro and micro environment as well as those of mediating and modifying variables;
- it must be able to account for inter-generational effects;
- it must be able to account for temporal effects;
- it must be able to account for cumulative effects; and
- it must be applicable in a range of settings (for example, less developed as well as developed countries) and to a range of health outcomes.

The framework draws on elements of the models discussed above. It could be used to guide study design, to contextualise a set of results or in the exploration of a single level. In my view, its main function is to place micro and macro environmental data in context and to emphasise causal relationships rather than the isolated effects of single causal agents. This is likely to lead to preventive and health promotive strategies which address the complexities of child health determination and avoid simple, 'quick-fix' solutions.

As an example, the decontextualised focus on maternal smoking as the main determinant of adverse pregnancy outcomes has led to failed health promotion

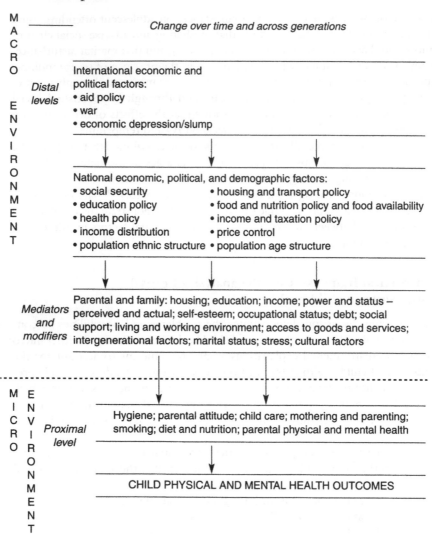

Figure 8.7 Theoretical causal framework for the determination of child health

interventions precisely because the micro level factor, in this case smoking, was isolated from the social context, macro level, in which it occurs. In the same way, it is unlikely that a desirable measure at the macro level, such as a ban on cigarette advertising, would have a short or even medium term effect on the level of maternal smoking in pregnancy. However, an overall strategy based on health and social policies which reduce the effects of material disadvantage and confront vested commercial interest, combined with micro level interventions aimed at building self-esteem and social support, is likely to be more effective in promoting child health.

Summary and conclusions

Child health determination cannot be studied, or interventions planned, by focusing exclusively either on the micro or the macro level. Researchers, health planners and policy makers need a framework within which to understand the complexities of the pathways between these levels. It might be argued that over emphasis on complexity may inhibit effective interventions and over complicate research design. As I hope this chapter has demonstrated, research designed to identify single causal agents and simple interventions based on their conclusions are likely to be ineffective precisely because they isolate variables and remove them from their context. Understanding of complexity should ensure that research findings, at whatever level, are set within causal pathways and chains allowing the conception and design of health interventions, based on these findings, to be sensitive to influences at micro and macro levels.

References

Aukett, A. and Wharton, B.A., 1995, Suboptimal nutrition, in: Lindstrom, B. and Spencer, N.J. (eds), *Social Paediatrics*, Oxford University Press, Oxford.

Bailey, P., Ong Tsui, A., Janowitz, B., Dominik, R. and Araujo, L., 1990, A study of infant mortality and causes of death in a rural north-east Brazilian community, *Journal of Biosocial Science*, 22: 349–363.

Baird, D., 1974, The epidemiology of low birth weight: changes in incidence in Aberdeen 1948–72, *Journal of Biosocial Science*, 7: 77–97.

Barker, D.J.P., 1992, *Foetal and infant origins of adult disease*, BMJ Publications Group, London.

Blackburn, C. and Graham, H., 1993, *Smoking amongst Working Class Mothers*, an information pack, Department of Applied Social Studies, University of Warwick, Coventry.

Browne, K., 1988, The nature of child abuse and neglect: an overview, in: Browne, K., Davies, C. and Stratton, P. (eds), *Early prediction and prevention of child abuse*, Wiley, Chichester.

Burns, J.L., 1947, Home and social environmental factors and the health of the school child, *Journal of the Royal Institute of Public Health*, 10: 325–336.

Butler, N.R., Goldstein, H. and Ross, E.M., 1972, Cigarette smoking in pregnancy: its influence on birth weight and perinatal mortality, *British Medical Journal*, ii: 127–30.

Caldwell, J., 1979, Education as a factor in mortality decline: an examination of Nigerian data, Professional Studies, 33: 395–413

Caldwell, J. and McDonald, P., 1982, Influence of maternal education in infant and child mortality: levels and causes, *Health Policy and Education*, 2: 251–267.

Chollet-Traquet, C., 1992, *Women and Tobacco*, World Health Organisation, Geneva.

Costello, A., Watson, F. and Woodward, D., 1994, Human face or human façade? Adjustment and the health of mothers and children, Centre for International Child Health, Institute of Child Health, London.

Danziger, S. and Stern, J., 1990, *The Causes and Consequences of Child Poverty in the United States*, Innocenti Occasional Papers, No. 10, Unicef, International Child Development Centre, Florence.

Davie, R., Butler, N. and Goldstein, H., 1972, *From Birth to Seven*, Longmans, London.

Douglas, J.W.B., 1951, Health and survival in different social classes: a national survey, *Lancet*, ii: 440–46.

Ebrahim, G.J., 1985, *Social and Community Paediatrics in Developing Countries*, Macmillan, London.

Evans, J-A. and Golding, J., 1992, Parental smoking and respiratory problems in childhood, in: Poswillo, D. and Alberman, E. (eds), *The Effects of Smoking on the Fetus, Neonate and Child*, Oxford Medical Publications, Oxford.

Farrington, D.P., Gallagher, B., Morley, L., Raymond, J. St. L. and West, D.J., 1986, Unemployment, school leaving and crime, *British Journal of Criminology*, 26: 335–56.

Fergusson, D.M., Horwood, L.J. and Shannon, F.T., 1984, Relationship of family life events, maternal depression and child-rearing problems, *Pediatrics*, 73: 773–6.

Finerman, R., 1994, 'Parental incompetence' and 'Selective neglect': blaming the victim in child survival, *Social Science and Medicine*, 40: 5–13.

Fogelman, K., 1980, Smoking in pregnancy and subsequent development of the child, *Child: care, health and development*, 6: 233–49.

Frisch, A.S., Kallen, D.J., Griffore, R.J. and Dolanski, E.A., 1992, Social variances and infant survival: a path analysis approach, *Journal of Biosocial Science*, 24: 175–183.

Garrett, P., Ng'andu, N. and Ferron, J., 1994, Poverty experiences of young children and the quality of their home environments, *Child Development*, 65: 331–345.

Golding, J. and Butler, N.R., 1984, The socioeconomic factor, in: Falk, F. (ed.), *The Prevention of Perinatal Mortality and Morbidity*, Vol. 3 of a series 'Child Health and Development', Manciaux, M. (series editor), Karger, Basel.

Golding, J., Fleming, P. and Parkes, S., 1992, Cot deaths and sleeping position campaigns, *Lancet*, 339: 743–9.

Graham, H., 1989, Women and smoking in the United Kingdom: the implications for health promotion, *Health Promotion*, 371–82.

Graham, H., 1993a, Smoking amongst working class women, *Primary Health Care*, 3(2): 15–16.

Graham, H., 1993b, *Hardship and Health in Women's Lives*, Harvester Wheatsheaf, Hemel Hempstead.

Grant, J., 1994, *State of the World's Children, 1994*, UNICEF with Oxford University Press, New York.

Harlap, S. and Davies, A., 1974, Infant admissions to hospital and maternal smoking, *Lancet*, i: 529–32.

Hart, N., 1993, Famine, maternal nutrition and infant mortality: a re-examination of the Dutch Hunger Winter, *Population Studies*, 47: 27–46.

Ineichen, B., 1993, *Homes and Health: How Housing and Health Interact*, E. & F.N. Spon, London.

Jolly, D.L., 1990, *The Impact of Adversity on Child Health – Poverty and Disadvantage*, Australian College of Paediatrics, Melbourne.

Keitner, G.I. and Miller, I.W., 1990, Family functioning and major depression: an overview, *American Journal of Psychiatry*, 147: 1128–37.

Kolvin, I., Miller, F.J.W., Fleeting, M. and Kolvin, P.A., 1988, Risk/protective factors for offending with particular reference to deprivation, in: Rutter, M. (ed.), *Studies of Psychosocial Risk: The Power of Longitudinal Data*, European Science Foundation, Cambridge University Press, Cambridge.

Kumar, V., 1993, *Poverty and Inequality in the UK: The Effects on Children*, National Children's Bureau, London.

Logan, S. and Spencer, N.J., 1996, Smoking, other health-related behaviour, socio-environmental context and the young, *Archives of Disease in Childhood*, 74: 174–9.

Lyons-Ruth, K., Connell, D.B. and Grunebaum, H.U., 1990, Infants at social risk: maternal depression and family support services as mediators of infant development and security of attachment, *Child Development*, 61: 85–98.

McKeown, T., 1976, *The Role of Medicine: Dream, Mirage or Nemesis?*, The Nuffield Hospitals Trust, London.

McLoone, P. and Boddy, F.A., 1994, Deprivation and mortality in Scotland 1981 and 1991, *British Medical Journal*, 309: 1465–70.

Madeley, R., Gillies, P.A., Power, F.L. and Symonds, E.M., 1989, Nottingham mothers stop smoking project – baseline survey of smoking in pregnancy, *Community Medicine*, 11: 124–30.

Martin, C., Platt, S.D. and Hunt, S.M., 1987, Housing conditions and ill health, *British Medical Journal*, 294: 1125–27.

Millard, A.V., 1985, Child mortality and economic variation among rural Mexican households, *Social Science and Medicine*, 20: 589–599.

Millard, A.V., 1994, A causal model of high rates of child mortality, *Social Science and Medicine*, 38: 253–68.

Miller, F.J.W., Court, S.D.M., Walton, W.S. and Knox, E.G., 1960, *Growing up in Newcastle upon Tyne*, Oxford University Press, Oxford.

Mitchell, E.A., Taylor, B.J., Ford, R.P. *et al.*, 1992, Four modifiable and other major risk factors for cot death: the New Zealand study, *Journal of Paediatrics and Child Health*, 28 (Supplement 1): S3–8.

Moy, R.J.D., Booth, I.W., Choto, R-G.A.B. and McNeish, A.S., 1991, Risk factors for high diarrhoea frequency: a study in rural Zimbabwe, *Transactions of the Royal Society of Tropical Medicine and Hygiene*, 85: 814–18.

National Children's Homes (NCH), 1992, *Deep in Debt: A Survey of Problems Faced by Low Income Families*, NCH, London.

Palloni, A., 1981, Mortality in Latin America: emerging patterns, *Population and Development Review*, 7(4): 623–49.

Paton, D.N. and Findlay, L., 1926, *Poverty, nutrition and growth: studies of child life in cities and rural districts of Scotland*, Medical Research Council, Special Report Series, No. 101, London.

Patterson, G.R. and Stouthamer-Loeber, M., 1984, The correlation of management practices and delinquency, *Child Development*, 55: 1299–1307.

Phillimore, P., Beattie, A. and Townsend, P., 1994, Widening health inequalities in northern England, *British Medical Journal*, 308: 1125–8.

Power, C., Manor, O. and Fox, J., 1991, *Health and Class: The Early Years*, Chapman & Hall, London.

Poswillo, D. and Alberman, E. (eds), 1992, *The Effects of Smoking on the Fetus, Neonate and Child*, Oxford Medical Publications, Oxford.

Reading, R., Openshaw, S. and Jarvis, S.N., 1990, Measuring child health inequalities using aggregations of enumeration districts, *Journal of Public Health Medicine*, 12: 160–67.

Roberts, H., Smith, S.J. and Bryce, C., 1995, *Children at risk? Safety as a social value*, Open University Press, Milton Keynes.

Rose, G., 1992, *The strategy of preventive medicine*, Oxford University Press, Oxford.

Rutter, M., 1988, Longitudinal data in the study of causal processes: some uses and pitfalls, in: Rutter, M. (ed.), *Studies of psychosocial risk: the power of longitudinal data*, European Science Foundation, Cambridge University Press.

Sampson, R.J. and Laub, J.H., 1994, Urban poverty and the family context of delinquency: a new look at structure and process in a classic study, *Child Development*, 65: 523–540.

Sanders, D., 1985, *The struggle for health*, Macmillan, London.

Save the Children Fund, 1985, *Towards a children's agenda: new challenges for social development*, Save the Children Fund, London.

Smith, D. and Nicholson, M., 1992, Poverty and ill health: controversies past and present, *Proceedings of the Royal College of Physicians*, Edinburgh, 305: 757–9.

Smith, G.D., Blane, D. and Bartley, M., 1994, Explanations of socio-economic differences in mortality: evidence from Britain and elsewhere, *European Journal of Public Health*, 4: 131–44.

Spence, J.C., Walton, W.S., Miller, F.J.W. and Court, S.D.M., 1954, *A thousand families in Newcastle upon Tyne*, Oxford University Press, London.

Spencer, N.J., 1996, *Poverty and child health*, Radcliffe Medical Press, Oxford and New York.

Spencer, N.J., Lewis, M.A. and Logan, S., 1993, Multiple admissions and deprivation, *Archives of Disease in Childhood*, 68: 760–62.

Spencer, N.J., Logan, S., Scholey, S. and Gentle, S., 1996, Bronchiolitis and deprivation, *Archives of Disease in Childhood*, 74: 50–52.

Szreter, S., 1988, The importance of social intervention in Britain's mortality decline, *c*.1850–1914: a re-interpretation of the role of public health, *Social History of Medicine*, 1: 1–37.

Townsend, P., Davidson, N. and Whitehead, M., 1992, *Inequalities in health: the Black Report and the Health Divide*, Penguin, Harmondsworth.

United Nations Development Program, 1994, *Human Development Report*, UNDP and Oxford University Press, New York.

Victora, C.G., Huttly, S.R.A., Barros, F.C., Lombardi, C. and Vaughan, J.P., 1992, Maternal education in relation to early and late child health outcomes: findings from a Brazilian cohort study, *Social Science and Medicine*, 34: 899–905.

Waxler, N.E., Morrison, B.M., Sirisena, W.M. *et al.*, 1995, Infant mortality in Sri Lankan households: a causal model, *Social Science and Medicine*, 20: 381–92.

Wedge, P. and Prosser, H., 1973, *Born to fail?*, Arrow Books, London.

Wilkinson, R., 1994, *Unfair Shares: The Effect of Widening Income Differences on the Welfare of the Young*, Barnados Publications, Ilford.

9 Healthy homes

Hossein Adibi

Introduction

In recent times, there has been an increasing interest in and need to view health in a holistic context. Today, more than ever in human history, we are extremely conscious and concerned about the state of human health.

The constitution of the World Health Organisation (1979) provides direction and strategy for the future. It broadly defines health as 'a state of complete physical, mental, and social well-being and not merely the absence of disease or infirmity'. This definition implies a more fundamental principle that 'The enjoyment of the highest attainable standard of health is one of the fundamental rights of every human being without distinction of race, religion, political belief, economic or social condition'. The establishment of such a right is the responsibility of government. This manifesto concludes that: 'Governments have a responsibility for the health of their peoples which can be fulfilled only by the provision of adequate health and social measures'.

Although this framework is quite clear, the dominant perspective of medicine continues to rely heavily on physical aspects of the body. It is well documented that the role of modern medicine in improving health and extending life has been exaggerated (Davis and George, 1993; Russell and Schofield, 1986).

The expansion of medical technologies, services and practitioner roles has undoubtedly been associated with significant health benefits for individuals. Effective surgery, antibiotics, and the relief of distressing symptoms during the course of an illness are primary among these. However, the contribution of medicine as a whole to the general health of the population has been oversold (Russell and Schofield, 1986: 146).

There is a substantial body of knowledge to support the idea that medical care (and even health care as conventionally defined) makes relatively little contribution to health compared with the potential contribution of broader social and cultural conditions.

A society has a clear obligation to protect its citizens, in so far as it is possible to do so, from environmental pollution, occupational hazards, infectious disease and other preventable causes of illness, dis-ease or disability. A society is also responsible for providing for basic human needs, such as food, shelter, clothing and satisfying work. The absence of such health sustaining circumstances inexorably leads to illness and dis-ease.

The question arises: what is the significance of 'home' in this context? This chapter explores some aspects of the relationship between homes and health and how it is possible for societies to move towards creating and sustaining healthy homes to the benefit of healthy human beings.

Characteristics of healthy homes

Throughout most of human history, individuals lived in a social world that was more or less unified and coherent. By contrast, modern life is typically segmented to a very high degree, so that the individual experiences a plurality of social worlds.

Modernity has liberated human beings from the narrow controls of family, clan, tribe or small community. It has opened up for the individual previously unheard of options and avenues of mobility. It has provided enormous power, both in the control of nature and in the management of human affairs. However, this liberation has had a high price: 'homelessness'.

Modern society is characterised by a large number of social relationships, most of them very superficial. The pluralistic structure of modern society has made the life of more and more individuals migratory and mobile. In everyday life individuals continuously alternate between highly discrepant social contexts. The individual migrates through a succession of widely divergent social worlds. Not only are an increasing number of individuals in a modern society uprooted from their original social milieu, but also no subsequent milieu succeeds in becoming truly 'home' either.

The 'homelessness' of modern social life has found its most devastating expression in the area of religion. Because of religious crisis in modern society, social 'homelessness' has become metaphysical. In Berger's words, social homelessness 'has become homelessness in the cosmos'.

Modern society's solution has been the creation of the private sphere as a distinctive and largely segregated sector of social life. A dichotomy in the individual's societal involvements has developed between the private and the public spheres.

One of the products of this dichotomy is the creation of the autonomous individual, and the consequence of this process is the individual's alienation from 'home'. Capitalism is perceived as a major fragmentary, 'alienating' and ultimately dehumanising, force which pits individuals against each other in merciless competitive conflict. 'Homelessness' is one extreme of this spectrum. What are the characteristics of healthy homes?

Healthy homes

Homes have been the units of our social, cultural, physical and environmental heritage for all of human existence. Living in human society, we often follow social rules because, as a result of socialisation, it has become habitual for us to do so.

Take, for example, the rules involved in language. Using language means knowing a variety of rules of grammar and speech. Most of the time, we simply utilise these without having to give them any thought, since we learned them in early childhood. It is only when we try to master a foreign language that we recognise how many rules have to be learned even to be able to speak simple sentences correctly.

Living in a home follows exactly the same rules. Through our socialisation we learn to develop a sense of belonging to where and what is called home. We take it for granted. Often we do not fully understand what it means to have a home, until faced with a situation where we have no place to live – when we have no home.

Our homes are not isolated physical units with autonomous management. Rather, they are extensions of social, economic, and managerial enclaves that exist in society. Our home economy, the ecology of our home, is affected by surrounding influential societal forces. The possibility of making our home healthy is determined by the availability of social resources.

Conceptually for this discussion, 'home' is considered to be a combination of physical (housing), environmental (neighbourhood), and social (family) factors. These characterising factors which greatly influence the state of people's health will now be explored in detail.

Housing

Good housing is fundamental for human well-being, and is a pre-requisite for human health and wellness. Important determinants of healthy housing include physical and social location, affordability and ownership, amenity and protection, and the provision of shelter and safety.

It turns out that 'cheap' high density housing areas that were, supposedly, rationally and effectively planned constitute and generate all kinds of problems: inhuman design, inappropriate materials of poor quality, social isolation, vandalism and other social problems.

It is a paradox that it is not only older, run-down houses that threaten people's health; many modern and renovated dwellings are similarly hazardous.

The physical standard of housing affects people's health. The layout of dwellings and residential areas and the relationships of dwellings influence the well-being of residents and their opportunities for social interaction.

The patterns of utilisation of dwellings modulates potential hazards for residents. Children and elderly people are especially at risk. Children must be protected from burns from electric and gas stoves and fireplaces, shocks and cuts from household appliances and poisoning from household chemicals. It is well documented that most home and leisure accidents occur during childhood and adolescence, and that they occur in or near the home. More than 46 per cent of the accidents happen in or around the home, almost 20 per cent in sports areas and 10 per cent in traffic (Bistrup, 1991). According to the European Home and Leisure Accident Surveillance System, home and leisure accidents in 1988 accounted for 75 per cent of accidents treated at accident and emergency departments (Bay-Nielsen, 1988).

Safe homes are expected to be accident-free homes. In British homes about 5,500 fatal accidents occur every year – some two-fifths of all fatal accidents. A further 2.2 million people have non-fatal accidents at home which need hospital treatment. Another 900,000 people are dealt with by general practitioners. The cost of domestic accidents to the health service in England and Wales is estimated to be £300 million (Graham, 1986).

Blackman and colleagues reported (1987) that accidents are the commonest cause of death in children aged over 1 year and account for one-third of all childhood deaths. About three-quarters of a million children are injured at home each year, and domestic accidents account for most of the falls, burns, and poisoning. Many of these accidents are preventable.

Another factor in housing affecting the health of residents is the building materials. Even today, many new houses pose a health risk to the inhabitants for this reason. There is evidence to indicate strongly that many new building materials, including coatings, epoxy varnishes and especially insulation, adversely affect the quality of indoor air, with consequences for human health which are not fully studied. As an example of biological pollution, house dust mites can cause asthma.

Overcrowding is another major problem affecting health. In the late nineteenth century tuberculosis was linked to poor urban housing, in particular to overcrowding and inadequate heating and ventilation.

Studies of the effects of housing on health are complicated by intersecting variables such as poverty, unemployment and social class. Although it may be difficult to prove that housing harms health as defined by a strictly medical model, yet if we accept the World Health Organisation's concept of health, the effects are obvious (Lowry, 1991).

The family

In discussing the social aspect of healthy homes, many social scientists have regarded the family as the cornerstone of society. It forms the basic unit of social organisation and it is difficult to imagine how human society could function without it. Although the composition of the family varies, such differences can be seen as variations on a basic theme. In general, therefore, the family has been seen as a universal social institution, an inevitable part of human society.

We have gone too far by replacing familistic relationships with contractual agreements in contemporary societies. Familistic relations are found among the members of a devoted family and among real friends. It is in these contacts that the individual ego is merged in the sense of 'we'. The individuals need one another, seek one another and gladly make sacrifices for one another. This is the noblest type of social relationship. In such a unity, a special contract becomes unlimited, all-embracing and all-forgiving. In today's society this is continuously diminishing and paving the way for increasing contractual relationships.

A contractual relationship is limited and specified. It does not cover the whole life of the parties, but only one narrow segment. The nature of contractual relationships is utilitarian and self-centred. It is based on the desire to obtain some advantage: pleasure, profit and the like. In this type of relationship each party remains individualistic and this does not lead to a homogeneous 'we'. A contractual relationship has the merit of constituting a voluntary agreement between free parties for their mutual benefit, but the relative proportions of these forms of relationship have changed in modern society.

The present is characterised by the golden age of contractualism, but society has failed to notice underground forces that have kept pace with the growth of contractualism. As a result, governments, not individuals, now decide, control and regulate all social relationships.

The decline of the contractual relationship in the family has taken the following forms:

- It manifests itself in a progressive disintegration of the contractual family of parents and children and of the circle of relatives.
- The disintegration shows itself in many forms: the tie binding husband and wife together, normally for life, has weakened enormously in the past few decades.
- The bond uniting parents and children has likewise become weaker: first, because of the increasing percentage of marriages without children; secondly, because children now leave home earlier than before; and thirdly because, when they are grown, they much less frequently continue to live with their parents and especially with their grandparents.
- The family has changed from an extended family into a nuclear family, resulting in the shrinkage of its size and functions.
- The family has lost many of its functions. In the past the family was the foremost educational agency for the young. Children now are withdrawn from the educational influence of the home at progressively younger ages. The place of the family is being taken by the nursery school, kindergarten, elementary school, high school, college and university.
- Until recently, the family was the principal place of socialisation for the newborn. At present this vital mission is performed less and less by the family. Increasingly the unstable family is a poor school for socialisation. The strong family creates a strong sense of moral and social integrity. Disintegrating families teach children lessons of moral laxity and loose relationships of antagonism and conflict between parents. Such a family cannot fail to create unstable people.
- The family today is less and less a religious agency. A number of other agencies perform this function.
- Other recreational functions of the family have likewise weakened or disappeared. Formerly the family took care of this. Now we go to the movies, theatres, night clubs, sport clubs and the like. Formerly the family was the principal agency for transforming one's psycho-social isolation and loneliness. Now families are small and soon scattered. Even when they work and live in one place, in the evening they disperse in quest of recreation. The result is that the family home becomes a mere 'overnight parking place'.

Accordingly, the family has lost most of its functions. It has shrunk in size, and has become increasingly unstable and fragile. Less and less is it able to provide comfort and joy which are the primary objectives of the contractual family.

Giddens (1991: 408) in describing the fundamental function of the family states that 'The home is in fact the most dangerous place in modern society. In statistical terms, a person of any age or of either sex is far more likely to be subject to physical attack in the home than on the street at night. One in four murders in the UK is committed by one family member against another.'

One major problem of the modern family is domestic violence. The emotional closeness of modern family life has its darker side: people who are intensely dependent upon one another are also very vulnerable. Estimates of the extent of family violence vary widely. For example, in US society, according to a 1985 national survey, about 16 per cent of couples reported at least one violent episode during the year (Gelles and Straus, 1988), and it is safe to assume that not all violence is reported. Although much domestic violence is directed against husbands and parents, the most common victims are wives and children (Brush, 1990). Reviews of the domestic violence research suggest that family violence is a complex response to the societal environment as well as to the internal dynamics of the family (e.g. Steinmetz, 1987).

Modern society is characterised by widespread inequality in socio-economic status, by gender, ethnicity and age. Inequality is a prime breeding ground for the use of excessive force by the powerful and for resentment among the less powerful and more dependent members. Under these circumstances it is not difficult to imagine how failure and experiences of frustration outside the home can be vented in the relative 'safety' of the household. Thus, despite claims that family violence touches all social classes, it is strongly associated with low educational and occupational status, early marriage, and unplanned pregnancy (Fergusson *et al.*, 1986). The absence of family violence is one of the major social characteristics of healthy homes.

Another major problem with the modern family is the alarming rate of divorce. It is estimated that in American society fewer than half of all marriages contracted today will remain intact for 30 years or more (Weed, 1989). Divorce has strong effects on the emotional well-being of children. Acock and Kiecolt (1989) indicate that the effects of divorce on children's emotional well-being appear to be stronger than those of parental death.

Hart (1976) argues that any explanation of marital breakdown must consider three major factors:

- the value attached to marriage;
- the degree of conflict between the spouses; and
- the opportunities for individuals to escape from marriage.

Some social scientists (Parsons and Bales, 1955; Fletcher, 1966) argue that the rise in marital breakdown stems largely from the fact that marriage is increasingly valued. Leach (1967) also suggests that the nuclear family suffers from an emotional overload which increases the level of conflict between its members. It can also be argued that the functional relationship between the family and the economic system, which involves the relative isolation of the nuclear family from extended kin, may have dysfunctional consequences for the family. It is in this context that Hart (1976) argues that the increasing divorce rate can be seen as a 'product of conflict between the changing economic system and the family as its social and ideological super-structure'.

In contemporary advanced industrial societies, there is an increasing demand for cheap female wage labour. Wives are encouraged to take up paid employment not only because of the demand for their services, but also because the capitalist controlled media have encouraged 'consumerism' and the demand for goods that families desire.

Conflict results from the contradiction between female wage labour and the normative expectations which surround married life. Working wives are still expected to be primarily responsible for housework and raising children, as well as to play a subservient role, to some degree, to the male head of the household. These contradict the wife's role as a wage earner since she is now sharing the economic burden with her husband. Conflict resulting from this contradiction can lead to marital breakdown.

Thus healthy homes should be conflict-free social units. The expectations of husband and wife should be based on the actual needs of the family and not on 'material aspiration and consumerism'.

Healthy homes require that we re-examine our value system within the family institution. Any fundamental change in this area requires changes in our attitudes towards our economic system, commercialism and consumerism. It seems that at present great emphasis is placed on privatisation and the undermining of public life. Consciousness raising of family members may be the major force for change in the societal value system which could bring the improvement that is made up of healthy homes.

Neighbourhood

The neighbourhood is considered to be an immediate extension of the home. Many activities, meaningful communications and interactions happen in neighbourhoods among the members. For its members, it provides a sense of identity and of belonging to a community. The neighbourhood also includes children's unstructured play areas and is a secondary unit of socialisation.

The environment outside the home plays a significant role in human health, social interaction and community life. The relationship between human health and the environment can be observed in many dimensions such as noise, vibration, chemicals and other types of pollution. Such dangers are all heightened by inadequate planning of the built environment as well as lack of concern for human needs, and non-availability of appropriate facilities for residence.

From the point of view of human health, industrial wastes are of considerable public concern. For more than half a century, industries in many countries have been disposing of poisonous chemicals at random. They have been discharging toxic substances into the atmosphere, local sewerage systems, streams and the environment. As an example there are at least 30,000 hazardous waste dump sites in the United States that pose a danger to water supplies (Macionis, 1991: 542). In addition, nuclear power plants produce waste materials that remain highly radioactive for hundreds of thousands of years.

With regard to social interaction, many neighbourhoods have never been developed on the bases of people's wants nor has their participation been invited. The result is 'efficient' urban renewal, but the consequence may be dwellings improved by the removal of tenants and the isolation of neighbourhoods. Such efficiency exemplifies physical, social, and aesthetic abuse of community and environment.

These unbreakable links between people and their neighbourhood constitute the basis for a broader approach to health. Such an ecological approach to the environmental and health requires a transcendence of narrow discipline bound perspectives.

Neighbourhoods are the framework for daily life and the foundation for love, vocation, community involvement and leisure. Yet as a consequence of rapid changes in attitudes towards consumerism and the market economy, neighbourhoods are experiencing serious difficulties. When traffic cuts through neighbourhoods, it exemplifies the physical and mental compartmentalisation of daily life. This contributes to the difficulties of neighbourhoods continuing to function as the framework for a whole life.

Differences in economic and social status seem to widen the gap between social classes and various segments of our society and our neighbourhoods. More and more people are isolated at home, and depend on social welfare payments. Statistics reveal that when we compare the number of jobless with the number of dwellings, at least one person is at home for 24 hours a day in more than half of dwellings.

Yet the people who have time to spend in their neighbourhood may have a great potential to effect change. Many ordinary employees and other people who think that stressful daily life, time pressure, separation of generations and the consumption patterns could be different may also want to participate in change.

In order to make our neighbourhood safe and healthy, there is a need for a fundamental change in our attitudes towards our neighbourhood and community. The fundamental requirement for this changes lies in informed active community participation. Engaging in projects for change requires awareness and motivation, but also a surplus of resources and energy.

Solving urgent problems in a neighbourhood so that it becomes a safe and attractive place for people to live can be successful only if solutions are found in a dialogue between the public authorities, tenants, clubs, local community and interest groups. The importance of such dialogue is that it inevitably crosses numerous local and sectoral boundaries.

It is obvious that creating supportive environments requires the political will to give neighbourhood residents influence in the planning and decision-making process, thus increasing the tenants' sense that the neighbourhood belongs to them and will help in establishing overall infrastructures that support social welfare and healthcare in neighbourhoods.

Emphasis should be placed on education, democracy and confronting the values of consumerism. Supportive environments for health must be quality environments, not in the sense of luxury but in the sense of caring for people.

Pollution must be stopped at the source. A sustainable solution is to use cleaner technology. Design of materials must be based on a conscious and environmentally friendly selection of raw materials. All processes involved should be non-polluting and without risks or dangers. The sustainable development of supportive environments for health requires increased inter-sectoral action and increased collaboration between the authorities and the general public.

Research done in the area of housing and neighbourhood development indicates that careful design, appropriate materials and respect for the human scale are necessary elements in our homes and environments. It is in this context that the health and well-being of individuals and communities should be examined and enhanced using architectural, economic and socio-psychological indicators, including the design, meaning and use of the built environment.

Enhancing health in housing, neighbourhood, and community environments requires integrating perspectives that include ecological, biological, environmental protection, resource protection and planning considerations with a perspective that emphasises human health and well-being, from a holistic context.

A strategy for linking health to housing and community environments should include supportive environments, sustainable development, ecological planning, inter-sectoral action, equitable access to the prerequisites for health, public participation, and community action; thus enabling people to increase their personal and community control of their lives.

In the physical context, low quality urban environments must be improved by ensuring access to light, fresh air, and to play and leisure areas. Noise, smoke, dust, and traffic reduction are other elements for consideration. Housing itself must be improved to an acceptable standard.

Community participation implies that citizens should be consulted in planning for the future, not only to promote democratic participation and give the people who are affected by planning an opportunity to voice their opinions, but also to persuade politicians and planners to find better planning solutions, for healthier neighbourhoods and cities.

Public participation implies legislative mandate that public opinion must be heard before planning decisions are made. Community advocacy, enablement, participation, and mediation – the central components of 'local management' – are crucial in establishing healthy neighbourhoods and health-supporting environments. A shift in values seems fundamentally necessary if we are to achieve supportive and sustainable environments for health. The potential of the public sector here is extremely important.

There is a need to integrate the physical planning of neighbourhoods and cities with health planning in order to create healthy environments. New and alternative approaches and methodologies are needed. These include the use of urban management and the efficient use of environmental health administration when working to create and enhance the quality of 'healthy homes'.

The extreme of homelessness

In addition to the above problems by which the health of individuals are at risk, there is a further massive social problem – 'homelessness'. This is a major issue affecting the health of increasingly large numbers of individuals in the contemporary world.

The extremity of 'homelessness' is used to include a very broad range of conditions and causes. It should be emphasised that homelessness rarely happens by choice. Our social and emotional development, the extent of parental care and guidance, our family structure, and early education are not only matters of individual or family choice. Rather, they are profoundly influenced by wider social forces which surround us.

Globally, more than 100 million people have no physical shelter (Sivard, 1988). In addition, more than one quarter of the world's population lives in inadequate shelter. In the Indian city of Bombay (now Mumbai), more than 100,000 people live, eat and sleep on the streets (Helmuth, 1989). It has been estimated that in cities in developing countries, one person in three lives in overcrowded, rented accommodation in slums. Their homes usually have no piped water, drains, electricity or refuse collection.

In the United States, in 1993, appropriations for low-income public housing fell from $30,200 million to $7,400 million. About 2.5 million units of low-cost housing were lost to conversion, abandonment, or demolition. Over one million single-room occupancy hotel units, which were commonly available to single adults, disappeared from the housing market between 1970 and 1982 (Ornstein and Schatz, 1993). In addition, in the early 1980s in the United States, social services for the poor were cut and eligibility criteria became further restricted. More than 37 million Americans are medically uninsured. A central and most serious consequence of this abandonment and neglect is the increase in ill health – in the context of 'healthy homes', an increasing number of homeless people. On any given night, in the United States, it is estimated that somewhere between 567,000 and 735,000 people are homeless. Annual increases are estimated to be between 15 per cent and 50 per cent. Men without children constitute the largest group of the homeless. Families with children are the fastest growing component.

According to a study of 1,500 homeless adults in the United States in 1993 by the Stanford University Centre for Research in Disease Prevention, two-thirds of the homeless were aged between 25 and 45. Only a few were elderly. The findings of this study are in contrast with studies conducted 30 years earlier when the number of homeless elderly was ten times larger.

Ornstein and Schatz (1993) in their study suggest that 60 per cent of homeless people were high school graduates, and almost 25 per cent had attended college. The study also showed that before these people became homeless, 78 per cent of the men and 43 per cent of the women stated that their primary source of income had been from their own jobs. Only 2 per cent of the men and 15 per cent of the women had been supported by public assistance. Most men had been employed in blue-collar jobs, while most women had worked in non-professional white-collar jobs.

Another striking fact concerning homelessness relates to child abuse. In the Stanford study, some 10 per cent of men and 16 per cent of women reported being removed from their parents and placed in foster care before the age of eighteen. Six per cent of the men and 29 per cent of the women had been sexually abused as children, and 13 per cent of the men and 26 per cent of women had been physically abused.

Our stereotypes of the homeless result from seeing those who manifest severe mental and physical health problems and are therefore the most 'visible'. These 'visible' homeless people have significant psychiatric and substance abuse problems. However, the most 'visible' do not account for the largest proportion of the homeless. Seventy per cent of women and 50 per cent of men surveyed mentioned no addictive or psychiatric disorders which affected them before they first became homeless. These are the 'invisible' homeless. We know little about them because their 'normal' appearance and behaviour allow them to blend in with the rest of the population.

Apart from their personal risk factors, virtually all homeless people are victims of societal forces and the economic problems of present societies. High levels of unemployment, coupled with high rates of inflation, reduce the buying power of average hourly earnings, especially for young workers.

Although 'being homeless' has been equated with lacking a dwelling, homelessness means not only that one does not have a dwelling, but also that one is unable to get or hold on to an available dwelling. Homelessness is closely associated with lacking the ability to 'live up to' the demands and expectations of society – that one should lead a socially stable life or at least be able to take care of oneself fully.

Homelessness is not a singular social problem, but is closely related to a total, or almost total, lack of living resources. Some homeless people have a dwelling, but because it is sub-standard or because they cannot stand being there, they live in the streets or in shelter, or stay with friends or in cheap residential hotels.

Homelessness has become a universal phenomenon. For example, in Denmark, the National Institute for Social Research estimated that about 20,000 people were homeless in 1981. In 1986 the number of people staying at reception centres, shelters and boarding houses was estimated to be between 20,000 and 23,000 (Hansen, 1986).

Homelessness is a social problem that is frequently associated with other health problems. Abuse of illegal and legal intoxicants and severe mental illness are frequent among homeless people. Homelessness among the mentally ill is growing especially fast: this seems to be an urban problem.

A three-month study in Denmark (Hansen, 1986) revealed that 10 per cent of patients admitted to a large mental hospital were homeless and that these homeless people were young. Most homeless people on discharge from hospital remain homeless. The authors of this study speculate that the closing of psychiatric institutions, consequent upon the establishment of district psychiatric care, has created a group of mentally ill people for whom no adequate and acceptable system of care now exists.

One study of homelessness (Robertson and Greenblatt, 1992) indicates that families were homeless because they were poor – too poor to be able to pay the increasing rental costs of housing. Most families in the sample had been poor long before they became homeless. Their lives seemed to progress from crisis to crisis rather than from week to week or month to month. Almost always the crises were connected with lack of money. Attempts to manage or to cope were influenced by the structure of the family and by the resourcefulness of mothers and their part-ners. Yet these efforts never seemed sufficient to overcome the basic lack of money. 'Solutions' were transitory, and 'success' was temporary; the crisis was unresolved and eventually the family became homeless.

The contemporary homeless population in the Western world, though diverse, tends to be distinguished from the general population by extreme poverty, low job skills, high unemployment rates, and a high incidence of personal-social adjust-ment difficulties, such as mental disorders. These families were disadvantaged in obtaining any social and health resources that could resolve their homeless status.

These characteristics have changed relatively little in recent times. Most notably, serious mental disorders have probably overtaken alcoholism as the single most common adjustment problem. Today, the homeless are more likely to be unem-ployed and are substantially poorer than their counterparts from the 1950s to the 1970s. Homeless people now tend to be younger and better educated. They also include more members of ethnic minorities and more women, both in absolute numbers and by percentage of population.

It is likely that upheavals in families of origin play a significant role in initiating homelessness. Marital separation and divorce, common in the backgrounds of homeless adults, are frequent precipitants. Research studies have found that low levels of current family contact and support exist among homeless populations. Underlying reasons include the absence of living family members, the rejection of the family by the homeless individual, and the family's intolerance and consequent rejection of the individual. The absence of social support – friends as well as fam-ilies – is central to homelessness.

Homelessness makes people ill. It is unhealthy for children, for their parents, for all human beings. In the extreme, homelessness is a fatal condition.

Conclusion

From earliest times, homes have been the bases of our social, cultural, physical and environmental heritage. Through the process of socialisation, we learnt to develop a sense of belonging to what and where is called home.

Home seems to be such a safe place. We picture it as a haven, a place to which we can retreat from the dangers of the world. We take it for granted. Often we do not fully understand what it means to have a home, until we are suddenly faced with a circumstance where we have no place to live – when we have no home.

People in modern times have suffered from a deepening sense of 'homeless-ness'. The impact of society on self has resulted in what might be called a metaphysical loss of 'home'. Homelessness has become a devastating experience for vast numbers of people in the modern world. For children, this loss comes at a point in their lives when the absence of a stable, nurturing setting is most injurious; when they are developing a sense of themselves and of their own identity.

It is vital that a practical plan be developed for the improvement of the conditions under which millions of people currently live in unhealthy homes.

This chapter confirms the view that our 'homes' are not isolated physical units of occupation, with the autonomy to manage them. Our homes, rather, are extensions of the social, economic and managerial enclaves that exist in today's society. The quality of our relationships at home is strongly influenced by larger social forces in society. Hence our opportunity to create healthy homes is determined by this availability of wider social resources.

Our homes do not function solely as physical shelters. Nor does the lack of domestic violence identify a healthy home. These factors constitute only part of the picture. Healthy homes, rather, are those in which a varied and diverse, dynamic and enriching long-term and meaningful relationship exists among the members. They share that unit, based on love and caring, which supports their developmental directions of self-actualisation and self-fulfilment.

If we take this definition as an 'ideal type' in Weberian terms, and take homelessness and living in slums and over-congested areas with no basic services for sustaining very basic day-to-day life, as the other end of the spectrum, then we can locate ourselves, as middle class, somewhere in between, struggling to reach that 'ideal type' of healthy home at the expense of global resources and its more deprived inhabitants.

Healthy homes will not come about merely from an increase in the living standard of the middle class in industrial societies. An enormous improvement in the quality of lives of millions of people who continue to live in deep poverty in our contemporary world community must be achieved.

The attainment of healthy homes requires an optimum balance between public and private life. In order to advocate and facilitate this change, it is fundamental that human consciousness and awareness be raised in the global context.

How this transition will take place, and whether we succeed or fail in achieving these aims, will be determined by the degree of our awareness and our level of appreciation of the quality of all human life. The attainment of this situation can only occur if the efficient and equitable management of our human and environmental resources is undertaken from within the context of these core human values.

Bibliography

Acock, A. and Kiecolt, K.J., 1989, Is it family structure or socioeconomic status?, *Social Forces*, 68: 553–571.

Bay-Nielsen, H., 1988, *European Home and Leisure Accident Surveillance System*, National Consumer Agency and National Board of Health, Copenhagen.

Berger, P., Berger, B. and Kellner, M., 1974, *The Homeless Mind: Modernisation and Consciousness*, Vintage Books Edition.

Bistrup, M.L., 1991, *Housing and Community Environments*, National Board of Health, Government of Denmark, Copenhagen.

Blackman, T., Evason, E., Melaugh, M. and Woods, R., 1987, *Housing and Health in West Belfast, A Case Study of Divis Flats and the Twinbrook Estate*, Belfast, Divis Joint Development Committee.

Bonnafe, D., 1990, *The Perception of Poverty in Europe in 1979*, Commission of the European Communities, Brussels.

Brush, L., 1990, Violence acts and injurious outcomes in married couples, *Gender and Society*, 4: 56–67.

Fergusson, D., Horwood, J., Kershaw, K.L. and Shannon, F.T., 1986, Factors associated with reports of wife assault in New Zealand, *Journal of Marriage and the Family*, 48: 407–412.

Fletcher, R., 1966, *The Family and Marriage in Britain*, Penguin Books.

Gelles, R. and Straus, M., 1988, *Intimate Violence*, Simon & Schuster, New York.

Giddens, A., 1991, *Sociology*, Polity Press, Cambridge, UK.

Graham, D., 1986, *The Divis Report – Set Them Free*, Divis Residents' Association, Belfast.

Hansen, F.F. (ed.), 1986, *The Poorest People in Denmark*, National Institute for Social Research, Copenhagen.

Hart, N., 1976, *When Marriage Ends: A Study in Status Passage*, Tavistock, London.

Helmuth, J.W., 1989, World hunger, amidst plenty, *USA Today*, Vol. 117, No. 2526, March, pp. 48–50.

Hess, B., Markson, E. and Stein, P., 1991, *Sociology*, Macmillan, New York.

Jensen, B.B. and Jorgensen, C.E., 1985, *Environment and Health*, Hans Reitzels Forlag, Copenhagen.

Koplev, K. (ed.), 1989, *Think Globally – Act Locally*, Arbejderbevagelsens International Forum, Copenhagen.

Leach, E.R., 1967, *A Runaway World?*, BBC Publications, London.

Lowry, S., 1991, *Housing and Health*, British Medical Journal, London.

Macionis, J., 1991, *Sociology*, Prentice Hall, New York.

McChesney, Y., 1992, Homeless Families, in: Robertson, M.J. and Greenblatt, M. (eds), *Homelessness: A National Perspective*, Plenum Press, New York.

Ornstein, B. and Schatz, H., 1993, *Homeless: Portraits of Americans in Hard Times*, Chronicle Books, California.

Parsons, T. and Bales, R., 1955, *Family, Socialisation, and Interaction Process*, Free Press, New York.

Raffestin, C. and Lawrence, R., 1990, An ecological perspective on housing, health and well-being, *Journal of Sociology and Social Welfare*, IXI: 143–160.

Robertson, M.J. and Greenblatt, M. (eds), 1992, *Homelessness: A National Perspective*, Plenum Press, New York.

Russell, C. and Schofield, A., 1986, *Where it Hurts*, Allen & Unwin, London.

Sivard, R.L., 1988, *World Military and Social Expenditure*, World Priority, Washington DC.

Stienmetz, S., 1987, Family violence: past, present and future, in: Sussman, M.B. and Steinmetz, S.K. (eds), *Handbook of Marriage and the Family*, pp. 725–764, Plenum Press, New York.

Weed, J., 1989, The life of a marriage, *American Demographic*, February.

World Commission on Environment and Development, 1987, *Our Common Future*, Oxford University Press.

World Health Organisation, 1979, *Measurement of Levels of Health*, World Health Organisation Publication, European Series No. 7.

Part III
Selected case studies

10 Health ecology and the biodiversity of natural medicine

Perspectives from traditional and complementary health systems

Gerard Bodeker

Abstract

Many of the approaches used in 'alternative' or 'complementary' medicine in North America, Western Europe and Australasia have their origins in long-standing traditional forms of health care from what are now known as developing countries.

According to the World Health Organisation, these traditions continue to serve the health needs of between 60 and 80 per cent of the population in their countries of origin.

International conservation organisations, such as the World Wide Fund for Nature (WWF) and the International Union for the Conservation of Nature (IUCN), have declared that the world's medicinal plant stocks are endangered through deforestation and over-harvesting.

In order to ensure sustainability in plant-based or natural health care strategies for future generations, policy makers will need to address the fundamental interaction between health and biodiversity. This will require new bilateral and multilateral investments in medicinal plant conservation and cultivation. It will also require strengthening – in some countries, creating – policies and an organisational infrastructure for the development of traditional medicine. Finally, it will require the generation of new and affordable clinical research methodologies along with national policies for ensuring the safety and efficacy of plant-based medicines.

An integrated approach to health and biodiversity planning is essential if sustainability is to be achieved in the traditional approaches to health care that continue to serve the majority of the population in developing countries and in the alternative approaches which serve a growing proportion of the population of industrialised nations.

Traditional health systems: policy, biodiversity and global inter-dependence

Many of the approaches used in 'alternative' or 'complementary' medicine in North America and Western Europe have their origins in long-standing traditional forms of health care from what are now known as developing countries. As the health strategies used in traditional health systems (for example Ayurveda and Traditional Chinese Medicine) have crossed national boundaries, they have come to be considered 'alternative' to the dominant health care system. However, according to the World Health Organisation, these traditions continue to serve the health needs of 60–80 per cent of the population in their countries of origin.

A growing number of developing countries have come to recognise that rising health care costs, compounded by the burden of the malaria and AIDS epidemics, point to the need for alternative approaches to meeting the basic health care needs of their populations. This has contributed to a resurgence of interest in traditional forms of health care and has led to a growth in research, new services and administrative programmes, and a recognition of the primary health care role of traditional health care practitioners.

At the same time, international conservation organisations, such as the World Wide Fund for Nature (WWF) and the International Union for the Conservation of Nature (IUCN), have declared that the world's medicinal plant stocks are endangered through deforestation and over-harvesting. The growth in the 'alternative' medicine market in industrialised countries is a significant contributor to this trend, with the global monetary value of plant-based pharmaceuticals in developed countries estimated to reach $500 billion by the year 2000 (Principe, 1991).

With the increase in public reliance on natural approaches to health care – be they 'alternative' or 'traditional', depending on the country of practice – they are likely to play an increasingly important role in meeting the health care needs of the majority of the world's population in the twenty-first century. In this event, serious national and international policy issues must be addressed in an integrated and comprehensive manner. This chapter attempts to outline the context and broad scope of the policy framework that will be needed if sustainability in natural health care is to be assured for future generations.

The global context

The World Health Organisation, which, as noted, estimates that the majority of the population of most developing countries relies on these traditional health approaches for their basic health care, has described traditional health systems as 'holistic' – 'i.e. that of viewing man in his totality within a wide ecological spectrum, and of emphasising the view that ill health or disease is brought about by an imbalance, or disequilibrium, of man in his total ecological system and not only by the causative agent and pathogenic evolution'. WHO documents describe traditional health systems as 'one of the surest means to achieve total health care coverage of the world population, using acceptable, safe, and economically feasible methods' (WHO, 1978).

The majority of the rural populations of developing countries have difficulty affording Western forms of health care. In cases of medical need, rural people may

Table 10.1 Relationship between traditional and modern medicine

Four broad organisational relationships have been observed between modern medicine and traditional medicine

Monopolistic	Modern medical doctors have the sole right to practise medicine
Tolerant	Traditional medical practitioners are not officially recognised but are free to practise on condition that they do not claim to be registered medical doctors
Parallel	Practitioners of both modern and traditional systems are officially recognised. They serve their patients through equal but separate systems, e.g. in India
Integrated	Modern and traditional medicine merged in medical education and jointly practised within a unique health service, e.g. China and Vietnam

Source: Stepan, 1983

have to travel for a day or more to reach a modern medical clinic or pharmacy. This results in lost wages, which is compounded by the cost of transport and the relatively high cost of the medicines themselves.

Traditional health systems in developing countries are typically the principal resort of the poorest levels of society when in need of health care. From the perspective of international health policy, they are of relevance since they are (i) locally available, (ii) sustainable, (iii) inexpensive, and (iv) potentially effective as a means of prevention, early intervention and self-medication and thus a potential source of cost reduction in health care.

Cultural factors play a significant role in the continued reliance on traditional medicine. Often villagers will seek symptomatic relief from modern medicine, while turning to traditional medicine for treatment of what may be perceived as the 'true cause of the condition' (Kleinman, 1983). Traditional medical knowledge is typically coded into household cooking practices, home remedies, and illness prevention and health maintenance beliefs and routines. Treatment is frequently a family based process, and the advice of family members or other significant members of a community has a major influence on health behaviour, including the type of treatment that is sought (Nichter, 1978).

Traditional and modern health care systems have interfaced with each other in four broad ways, as shown in Table 10.1.

Asia

Traditional health systems in Asia have been incorporated as formal components of national health care for several decades. After more than a century of grappling with the relationship between Western and traditional health care systems, India gave an official place to the Ayurvedic and Unani medical systems through the Indian Medicine Central Council Act of 1970. In 1995, a new Department of

Indian Systems of Medicine was established, with a permanent secretary and a plan for a cabinet minister responsible for the department. There are now more than 200,000 registered traditional medical practitioners in India. While traditional methods of training continue according to the *guru/chela* (master/apprentice) relationship, the majority of Indian practitioners have received their training in degree-granting government colleges of Ayurvedic or Unani medicine.

China has had a policy of integrating traditional medicine into national health care for more than three decades and has an extensive and integrated national programme in which modern and traditional medicine are combined as formal components of national health care provision. In both India and China, the traditional health sector provides the majority of health care to the poor and to rural communities.

Pakistan and Bangladesh both have departments of traditional medicine within their Health Ministries. In Sri Lanka, there is widespread use of the Ayurvedic system of medicine, with a Department of Ayurveda within the Health Ministry. In Nepal, there is a college of Ayurvedic education in Kathmandu and a Department of Ayurveda within the Health Ministry.

In Korea, where health insurance coverage is available for Oriental medical treatments, between 15 per cent and 20 per cent of the national health budget is directed to traditional medical services. Korean government reports indicate that traditional medicine is favoured equally by all levels of society (Choe Won Sok, 1995). In Japan, where physicians have been permitted to both prescribe and dispense medications, over two-thirds of all physicians are reported to prescribe herbal medication at times, some with great frequency (Norbeck and Lock, 1987).

The Americas

A number of South American countries have established departments or divisions of traditional medicine within their health ministries. In Mexico, whose population includes 9 million indigenous people, there has been a comprehensive programme to revitalise the indigenous health traditions. A programme of ethnomedical and pharmacognostic research has identified over a thousand traditional medicines. The government and indigenous organisations have established training centres to communicate traditional medical knowledge to new generations of health care workers. Traditional health care practitioners in Cuetzalan, where an extensive integrated community health programme has been implemented, have been examining ways to cultivate medicinal plants under controlled conditions as a means of generating income for their association (Zolla, 1993).

Hospitals of traditional medicine have been established in a number of rural areas and the Mexican Constitution is reportedly under revision to give traditional forms of health care a constitutional place in national health care provision (Argueta, 1995).

During the war in Nicaragua, the international economic blockade resulted in a severe shortage of medical and pharmaceutical supplies. As a matter of necessity,

Nicaragua in 1985 began to re-appraise its herbal traditions as a potential means of addressing national medical needs. The government established a new department of indigenous medicine to develop 'popular and traditional medicine as a strategy in the search for a self-determined response to a difficult economic, military and political situation' (Sotomayor Castellon, 1992).

This new department undertook a programme of ethnobotanical research in which more than 20,000 people throughout Nicaragua were interviewed concerning their use of traditional herbal remedies, their methods for preparing these, and the locations and plants from which the *materia medica* were derived.

In Canada, the Midwifery Act (Bill 56) of 1994 formally exempted aboriginal midwives from its jurisdiction and regulatory process. This has given new impetus to the revitalisation of traditional midwifery in Canada. Even prior to this legislation, the Inuit Women's Association of Canada had developed a programme to revitalise traditional birth practices. Video recordings of women who were midwives for many years in their own communities are used to train young Inuit women in the use of traditional delivery methods in combination with Western training in hygiene and safety (Flaherty, 1993). The Six Nations Aboriginal Traditional Midwifery Training Program is another instance of revitalisation of local midwifery traditions where traditional approaches are emphasised in the delivery process.

In the United States, native American communities have been incorporating traditional forms of treatment into Indian Health Service (IHS) alcohol rehabilitation programmes. An analysis of 190 IHS contract programmes found that 50 per cent offered a traditional sweat lodge at their site or encouraged its use. Treatment outcomes were better for alcoholic patients when a sweat lodge was available. In addition, the presence of medicine men or healers greatly improved the outcome when used in combination with the sweat lodge (Hall, 1986).

Africa

A number of African countries have developed programmes within their health ministries to support the study and promotion of traditional health care within primary health care services. In addition, research centres have been established in many countries and a number of regional collaborations exist among natural products researchers, including the Natural Products Research Organisation of East and Central Africa (NAPRECA) and the Medicinal Plants Group of the Organisation of African Unity.

In Africa, governments presented with overwhelming drug bills in the face of the growing AIDS crisis are looking to their indigenous medical traditions and medicinal plants to identify inexpensive and effective treatments at least to alleviate the suffering of AIDS victims (Kabatesi, 1998).

In Uganda, for example, where 70 per cent of trained medical doctors emigrate on graduating, there is inevitably continued strong reliance on traditional sources of health care. By contrast to doctor/population ratios of between 1:20,000 and 1:200,000, the population is served by approximately one traditional practitioner per 200 people (Mubiru, 1994). In articulating national priorities in drug research, the Ugandan National Drug Policy and Authority Statute (1993) has stated that the national drug policy shall 'intensify research in all types of drugs including traditional medicines'. The Joint Clinical Research Centre in Kampala and the Uganda AIDS Commission in conjunction with traditional healers' associations have investigated several traditional treatments for opportunistic infections associated with HIV and AIDS. A February 1994 newspaper report quotes an official of the AIDS Commission as saying that recent research had shown that 'traditional medicine is better suited to the treatment of some AIDS symptoms such as herpes zoster, chronic diarrhoea, shingles and weight loss' (Kogozi, 1994).

In Malawi, traditional healers have received training in counselling people in AIDS prevention and care and are considered a significant component of the national AIDS control programme (Berger, 1995).

In 1990, the World Health Organisation held a meeting on 'Traditional Medicine and AIDS' which resulted in the elaboration of guidelines and protocols for clinical evaluation of the safety and efficacy of traditional remedies (WHO, 1990). A conference of UNAIDS in 1998 reported on a growth in collaboration with traditional healers in the fight against AIDS in Africa.

Health care costs and traditional health services

Typically more than 80 per cent of health budgets in developing countries is directed to services that reach approximately 20 per cent of the population. Of this, around 30 per cent of the total health budget is spent on the national pharmaceutical bill.

The cost of drugs and the emergence of new epidemics are contributing to a resurgence of interest in traditional medicine on the part of governments. With approximately 30 per cent of the health budgets of developing countries being directed to the cost of drugs produced in industrialised countries, the prospect of dealing with epidemics such as AIDS, and the new rise in the incidence of malaria and tuberculosis, is forcing many governments to look to their indigenous systems of medicine and medicinal flora for low-cost solutions (Bodeker, 1998).

Acutely inadequate funding for research and service delivery in traditional medicine has been a long-standing barrier to progress in these areas. Conventional development funding programmes have favoured the modern medical sector exclusively, even though the majority of the population in most developing countries use traditional forms of health care. Revenues from patient fees are low, since traditional health care services are typically the first resort of the poor.

From the perspective of developing country health economics, medicinal plants stand as more viable and direct sources of medicine since the cost of producing synthetic drugs from natural products is reportedly as much as ten times that for direct extraction from medicinal plants (Principe, 1991).

Research policy in traditional health care

Biodiversity prospecting and developing country health care

In the 1990s, the term 'biodiversity prospecting' has come to be used to describe the search for potentially bioactive agents in plants or mixtures used by traditional medical practitioners (Reid *et al.*, 1993). There has been widespread popular press coverage for the research activities of companies engaged in biodiversity prospecting, and the public message conveyed has been that indigenous medical knowledge offers the promise of leading to the discovery of new drugs which will benefit mankind. The assumption is that these drugs will be synthetically produced models of single bioactive agents identified through screening tests conducted on medicinally important plants or mixtures of plants.

The National Institutes of Health in the United States have initiated two drug discovery projects along these lines – one in 1992 from the National Cancer Institute and the other in 1993 from the National Institute of Allergies and Infectious Diseases. The objective here is to link drug discovery with conservation initiatives.

Typically, practitioners of traditional medicine view the biomedical search for a single, so-called 'active ingredient' as an inappropriate application of their empirically and culturally grounded health knowledge. It is often pointed out that traditional medicine avoids the use of a single ingredient or extract and uses a complex mixture. The chemicals in the different plant ingredients may serve variously to offset side effects in other plants in the mixture, to increase cellular uptake of the chemicals which address the pathology, to stimulate a generalised immune response so that it is not a single receptor site that is being targeted but rather a systemic healing response that is being activated. In addition, extra herbs are added to the mixture to address the specific imbalances and medical needs of the presenting patient. This pharmacological model is complex and sophisticated and calls for a synergistic approach rather than the reductionistic 'active ingredient' approach to drug development.

Duke has characterised this view as follows: 'The synergic shotgun can therefore be better for treating pathogens, of humans, animals, and plants, than the solitary, synthetic, "magic bullet", so often patterned after just one of the natural compounds, wasting all the others and their potential, often provable synergy' (Duke, 1995). Indeed, research shows that, in the case of malaria at least, the traditional approach is more effective. The strains of malaria that have developed resistance to chloroquine and mefloquin are still not resistant to the original cinchona bark – the natural source of quinine, on which synthetic anti-malarial drugs were modelled (Bodeker *et al.*, 1998).

From a health policy perspective, national and international research strategies which view traditional medicine as raw material from which other, synthetic products can be developed risk skewing the national research endeavour away from the study of traditional health systems as valid health care services systems in their own right. Rather, such an approach favours the development of drugs from Third World materials to serve, not the developing country health priorities such as malaria, schistosomiasis, diarrhoeal disease and malnutrition, but First World health concerns such as cardiovascular disease and cancer.

Research methodologies

Research data gathered on traditional medicine in most countries tend to be in the form of pharmalogical studies on individual medicinal plants. Clinical trials designed to determine clinical outcomes of traditional mixtures of medicinal plants are rare (Parry, 1997).

In generating national research agendas that are appropriate to the evaluation of traditional health strategies, countries and international development agencies need to recognise that there are research methodologies suited to the evaluation of complementary medical modalities (Lewith and Aldridge, 1993). In order to evaluate the cost-effectiveness of traditional medicine and, eventually, to design a cost-effective intervention strategy, traditional and biomedical strategies might be compared separately as well as in combination. Such an approach would evaluate the clinical methods, the cost and the effectiveness of each system and the combined systems in the treatment of major diseases.

An outcomes measures methodology could be employed to address the effect of composite modalities in one or other medical system (Pelletier, 1991). If a traditional treatment for diarrhoea (e.g. a mixture containing the leaf and/or the green fruit of guava; local herbs) or for malaria (e.g. *tulsi* or *neem* in India, in Vietnam the herb known as '*tung son*', in a mixture with other anti-malarial herbs) were found to be equally effective as modern treatments and cost one-tenth to one-hundredth or less of the cost of the modern treatment, the case for traditional medicine would be clarified. If modern treatments were more effective, then this would clarify the importance of using modern treatments with that particular condition. Such information could assist in making choices about the use of modern or traditional treatments for different conditions. If a combination of treatments from both systems were found to be most effective, guidelines for combining could be developed and the economics of combination could be determined as the basis for developing national strategies.

Safety issues

Concerns are often raised by international development agencies and even developing country health ministries on safety issues pertaining to traditional health practices. Certainly, care needs to be taken to ensure that adequate safety data are available before health ministries recommend widespread use of specific herbal preparations. However, this need not be a lengthy or costly process if simple toxicological testing procedures are implemented at the national and local levels (Roy Chaudhury, 1993). Training in safe delivery practices and hygiene are already being undertaken in many countries with traditional birth attendants, as for example the case of the Inuit Women's Association in Canada. To place the issue of potential toxicity of herbal medicines in perspective, in the United States where one in three people report using some form of 'alternative' medicine (Eisenberg *et al.*, 1998) and where the alternative medicine industry is valued at $28 billion per year, plant poisonings are almost exclusively due to consumption of toxic ornamental plants, not herbs. Such poisonings resulted in only one fatality. In the same year, fatal (non-suicide) poisonings by anti-depressants, analgesics, sedatives and heart drugs totalled 414 (Fugh-Berman, 1993).

Constraints to research

There are a number of constraints to the research capabilities of universities, hospitals and institutes in most developing countries. A common situation is the acute shortage of modern, reliable scientific and diagnostic equipment.

Another constraint to the conduct of research is the professional isolation that traditional medicine researchers have experienced. While biomedical scientists may have studied overseas and regularly attended international scientific meetings, due to the marginal status of traditional medicine at the national policy level researchers investigating traditional medicine find little funding support either for their research or to participate in international scientific exchanges.

Scientists working in developing countries in the traditional medicine field generally fall into one of two main categories: those conducting pharmacognostic research on the medical properties of plants and those evaluating the diagnostic and treatment efficacy of the various clinical modalities of traditional medicine. In most countries, pharmacognostic research is more prevalent and methodologically sophisticated than clinical research, reflecting the extent to which the pharmacological model has dominated evaluation of traditional medicine.

Often, due to technical limitations, much of the research is not of a standard to be acceptable to international publications. Clinical research may lack methodological sophistication and statistical expertise. Scientists in the traditional health field have minimal access to electronic data bases and hence reduced access to current literature in their field. A concept for an international data base has been proposed in which an international collaboration would link alternative and traditional medical researchers in different parts of the globe (Wootton, 1995). Such international collaborations in developing electronic data bases on the world's traditional medical knowledge would assist these scientists and would benefit from the data available in research institutes in their region.

Most developing country scientists see the need for extensive collaboration with researchers in other countries, where methodology, equipment and experience is more advanced. They also recognise the importance of regular participation in the international scientific community, through conferences, meetings and sabbaticals if their research is to advance. There is also active interest in having international scholars spend periods of time in traditional medicine institutions in developing countries, assisting in the development of local research agendas, studies and capabilities.

Finally, there is strong interest in attracting support from international research foundations, which typically fund exclusively biomedical research and development. The urgency of this felt need is underscored in the last four words of the title of a 1993 publication from the Vietnamese National Institute of Burns, which utilises traditional medicine in the treatment of serious burn injury: 'Establishment of New Scientific Centre of Viet Nam – The National Institute of Burns, Named Le Huu Trac – That Needs Much Support.' (Le and Bodeker, 1997)

Biodiversity

> 'The forest makes no demands for its sustenance and extends protection to all beings, offering shade even to the axe man who destroys it.'
>
> Buddha – from a sign in a pine forest in Bhutan (cited in McNeely, 1993)

As noted in the section on Research policy, there has been a recent growth of interest in traditional medicine from the international pharmaceutical industry as well as from the natural products/alternative medicine industry in Western Europe and the United States.

Traditional medicine has come to be viewed by the pharmaceutical industry as a source of 'qualified leads' in the identification of bioactive agents which can then be synthetically modelled for the production of patented, modern drugs. The growth of interest in this area is evidenced by the fact that whereas, in 1980, very few international companies were and 'no U.S. company was working on higher plants' (Lewington, 1993), by 1990, 223 companies worldwide were investigating medicinal plants as the source of new drug leads (Fellows, 1991).

The international herbal medicine industry relies on medicinal plants from developing countries – particularly India and China – to provide a significant portion of the raw materials used in herbal medicine production.

Concern has been expressed by traditional medicine organisations in developing countries that this trend does not contribute to the development of traditional medicine as a health care system for the rural communities and the poor, but, rather, is designed to serve the demand for new drugs in industrial countries. In addition, the conservation benefits are considered to be modest and concern over intellectual property rights of traditional medical custodians has emerged as an important issue in the debate surrounding this trend (See UNDP, 1994, FAO, 1997).

Intellectual property rights

A widely cited example of biodiversity prospectors forming contractual agreements which recognise developing country intellectual property rights is the agreement between Merck Pharmaceuticals and INBio, a Costa Rican taxonomical research centre (Reid *et al.*, 1993). Under this agreement, Merck granted $1 million to INBio to gather species of tropical plants, insects and soil samples. At the same time, Merck was spending $1 billion a year on research, and critics have asserted that Merck 'wrung a good $10 million worth of good press out of a mere $1 million gamble' (Joyce, 1994). Other critics have argued that this agreement does not accord any rights to local herbalists or to the community but to the scientific organisation that stores and classifies the herbs (Dutfield, 1997). The approach has been rejected as a model for development by indigenous organisations, including the World Council of Indigenous Peoples.

Wade Davis, writing in the American Botanical Council's publication, *Herbal Gram*, quotes Joyce (1994): 'To transform a drug into a commercial product takes on the average 10 years and $350 million. Once the development costs are discounted, and assuming that the product nets annually $10 million, a royalty rate of 3 per cent would yield approximately $30,000 a year (p. 66).' At this rate the promise of a green goldmine awaiting indigenous communities involved in biodiversity prospecting seems more of a mirage than a prospect.

In the face of increased interest in their medicinal knowledge, some healers have been forced into deceptive practice as a means of protecting what is effectively their trade secret. In writing on Ethiopian government policy on gathering information on medicinal plants, Bishaw has noted that: 'Some of the healers interviewed by the writer claim to have complied with the Coordinating Office's demands, but only by submitting either a partial or a wrong list of the plants, animal parts, and minerals they would actually use in treating patients. They defended their actions by arguing that no-one had the right to take away an important means of their livelihood without any assurance of involving them both in the testing of their drugs and in the sharing of the possible benefits should these medicines be found useful enough for mass production and marketing' (Bishaw, 1991).

In another instance involving Ethiopia, Toledo University in Ohio, has been awarded a patent on the application of the indigenous Ethiopian moluscicide, *Phytolacca dodecandra*, commonly known as Endod. Endod is used for a variety of medicinal purposes in East Africa and the berries of the plant have been used traditionally as a soap in Ethiopia. Research in Ethiopia has found the berries to be an effective and inexpensive moluscicide providing protection against the water-borne snails which carry the larvae of the schistosome parasite which cause schistosomiasis. Research at Toledo University has examined the effect of Endod against the zebra mussels that are plaguing the Great Lakes and found that Endod is very effective against these molluscs. The Toledo patent on Endod as used against the zebra mussel is a source of controversy in Ethiopia. The question has been raised as to who owns the knowledge: scientists, the traditional custodians of Ethiopian medicinal knowledge, the people of Ethiopia or an American university (Seeds of Survival, 1993; RAFI, 1993, Down to Earth, 1993).

Shrinking supply and increased demand for medicinal plants

'The spirits of our ancestors have warned us of calamity should our sacred groves be destroyed.'

Elders of Chale Island, Kenya (Wilson, 1993)

In a 1988 document referred to as the Chiang Mai Declaration, the International Union for the Conservation of Nature (IUCN), the World Wide Fund for Nature (WWF) and the World Health Organisation referred to the critical situation with regard to loss of medicinal plants through over-harvesting and habitat destruction. The Chiang Mai Declaration noted that the loss of certain medicinal plant species and reduced supply of other important plants would have a direct impact on human health and well-being, as these plants form the basis of the medicines used by the rural majority of most developing countries. The Declaration drew attention to the urgent need for international cooperation and coordination to establish medicinal plant conservation programmes in order to ensure that adequate supplies are available to future generations (Akerele *et al.*, 1991).

A 1993 Unesco and WWF report notes that 'there is significant evidence to show that the supply of plants for traditional medicine is failing to satisfy demand' (Cunningham, 1993). The problem is reported to have been exacerbated by three main factors:

- There has been an increase in demand due to population growth and urban expansion. Paradoxically, urbanisation leads to an increased rather than a decreased demand for traditional health services. This may be due to an attempt to preserve cultural traditions. It may also reflect the inability of people to self-medicate, as is the case in rural areas, from plant sources growing in their local area.
- Medicinal plant harvesting has shifted from being the specialised activity of traditional medical practitioners. The new situation is one in which there is an informal sector of commercial plant harvesters whose primary motivation is profit.
- Changes in land use for forestry, agriculture, fuel supply, etc. has resulted in a decline in the total area of natural vegetation available as a source of medicines.

The international trade in medicinal plants, particularly the growth of demand from the European and United States herbal medicine markets, has brought new pressure to bear on the wild stocks of medicinal plants in developing countries. As noted in the introduction to this article, the global monetary value of plant-based pharmaceuticals in OECD countries is estimated to reach $500 billion by the year 2000 (Principe, 1991). In a 1982 study, India alone was found to supply 12 per cent of the European market. India in turn is estimated to gather 90 per cent of its medicinal plants from wild sources. China, which uses 700 million tons of medicinal plants in its domestic market each year, is estimated to gather at least 80 per cent of its medicinal plants from wild sources (He, 1997).

The German herbal market, which accounts for 70 per cent of the European market, was valued in 1989 at US$1.7 billion (Lewington, 1993). Traders in Hamburg supply both the German herbal market and other companies in Europe. TRAFFIC, a WWF-sponsored organisation which monitors trade in endangered species, has reported that traders typically form direct contractual relations with 'growers' (i.e. harvesters) and avoid working with middlemen. In general, medicinal plants are reported to be traded in the absence of tariff restrictions, since most plants and crude drugs are exempt from duty (Lewington, 1993). A case in point is Nepal where, it is reported, 'Hundreds of varieties of herbs in all incarnations – leaves, roots, stems, extracts – continue their journey from remote crags to staging posts in the hills and then to the Tarai. Through a time-tested network of legal and illegal routes, the bundles and sacks are heaved onto trucks, they hop on international flights, board trains and find berths in cargo vessels. Some are bought up by the Ayurvedic giants in India and others are acquired by cosmetics firms abroad, while perhaps the highest value usage is by pharmaceutical multinationals and their research laboratories in Europe and America' (Aryal, 1993).

Customs officers are reportedly bribed and allow trucks to pass across the border unchecked. Himalayan herbs make their way into India from Nepal, which not only loses its medicinal plant biodiversity through over-harvesting, but receives no revenue from the sale of its national resource and is not acknowledged as the source of the plants when they are re-sold by Indian companies to the European herb trade. Indeed, European medicinal plants are also under severe pressure (Lange, 1998).

TRAFFIC found that 'The point of view of UK traders appears to be that any conservation considerations (specifically, cultivation as opposed to wild harvesting) are largely an unaffordable luxury at the moment' (p. 29). TRAFFIC considers that the prerequisite for change is the creation of a commercial environment which reduces the constraints on herbal medicines that are prevalent in Europe and generates growth in the industry sufficient to satisfy buoyant consumer demand.

However, motivation to ensure future supplies is called into question by the actions of some companies. WWF and Unesco have jointly reported on the loss of *Prunus africana* in Cameroon due to pressure of over-harvesting to meet the European market for a drug used in the treatment of benign prostatic hypertrophy (Cunningham and Mbenkum, 1993). With *Prunus africana* now listed as an endangered species in Cameroon due to over-harvesting by these companies, the companies concerned have turned to other African countries for their sources of supply.

In simple economic terms, demand is increasing and supply is decreasing. Despite the call in the Chiang Mai Declaration for urgent international action, the supply side of traditional medicine (i.e. medicinal plants and their continued availability) has been almost completely neglected at the policy level. The net impact of this situation of unsustainable usage patterns is the potential loss of important indigenous *materia medica* and a resultant lack of availability of the medicines that currently serve the health needs of the greater majority of the world's population.

Many indigenous and conservation organisations are concerned about these trends. WWF has developed guidelines for countries to protect themselves and their financial interests in an increasingly competitive international market in plant-related pharmaceutical products. The Biodiversity Support Programme (BSP), a USAID-supported consortium involving WWF, the Nature Conservancy and the World Resources Institute, has issued guidelines for the sustainable harvest of non-timber plant resources in tropical forests (Peters, 1994). Six basic steps are outlined in the BSP strategy:

1   species selection – i.e. medicinal plants or specific types of medicinal plants. The choice of plants will be based on economic, social and sustainability criteria;
2   forest inventory – baseline data are needed on the extent and utilisation patterns of the medicinal flora and fauna in order to be able to monitor changes in wild stocks. BSP favours the use of the Geographical Information System (GIS) as a biodiversity monitoring system.
3   yield studies – local collectors would be trained to weigh, count, or measure the quantity of medicinal plants or plant products such as nuts, seed pods, etc.
4   regeneration surveys – every five years, regeneration studies are recommended to track the fluctuations in crop density due to harvesting patterns;
5   harvest assessments – quick, visual appraisals to detect growth problems;
6   harvest adjustments – regulation of the number or size of plants being exploited and regulation of the total area from which the resource can be harvested.

However, conservation alone will be insufficient as a means to address the supply side of 'traditional' and 'alternative' medicine. Active cultivation programmes will also be needed, and trade policies which introduce accountability regarding the sustainability of medicinal plants used in herbal medicines are also of central importance in ensuring supply.

Organisations such as the United Nations Food and Agriculture Organisation (FAO) should support development of appropriate agricultural methods for cultivating medicinal plants to serve local health care needs as well as to meet the demands of export markets. Agricultural questions need to be considered: such as the issue of whether it is economical to invest in large scale cultivation of medicinal plants as an alternative to continued reliance on wild sources of supply, and appropriate and inappropriate cultivation methods. Plant genetic research centres in India, Ethiopia and elsewhere have been working on these issues, and their experience should be combined with that of local NGOs such as the Foundation for Revitalisation of Local Health Traditions, based in Bangalore, India, and the Ugandan NGO, Rukararwe Bumetha, both of which combine traditional health care revitalisation services with biodiversity conservation strategies.

Economic analyses are needed to determine the role of traditional medicine in contributing to economic development – as a low cost medical alternative, as a source of income for families and farmers, as a means of attracting foreign investment, particularly from neighbouring countries as well as from the European and US herb trade. And international trade policies in herbal medicines need to be linked with biodiversity conservation, with companies taking responsibility for reporting on the source and replenishment of the herbs used in their herbal medicines. Perhaps the labels 'Not tested on animals' and 'Organically grown' will be joined by a herbal medicine label which reads 'Sustainably harvested in (name of country)'.

A paradigm shift

In conclusion, the shift in thinking that has been occurring with respect to traditional health systems reflects a recognition of their importance in health care

Table 10.2 Old and new perspectives on traditional systems of health care

Old	New
Primitive	Holistic
Ineffective	Cost-effective
Marginalised	Locally available
Becoming extinct	Undergoing renewal
Needs to be regulated	Needs to be promoted
Source of leads	Valid in its own right
For pharmaceutical industry	With local economic value
'Active ingredient'	Synergistic activity
Unlimited supply	Sustainable harvesting basis of future supply

delivery and of their potential role in providing new solutions to the significant health challenges faced by humanity at the end of the twentieth century.

Table 10.2 outlines the components of this shift in perspective (Bodeker, 1994).

Conclusions

A new role for traditional health systems requires that they be included as a matter of policy in planning and budgeting for health care. National and international planning in this field should be done in consultation with traditional medical practitioners and their representative organisations.

Priorities that should appear on national and international policy agendas for traditional health care are:

- *Legislation* that protects and strengthens traditional health knowledge and public access to safe and effective traditional health services.
- *Research strategies and institutional development* designed to evaluate traditional health care modalities in a way which is consonant with their paradigmatic framework and which has direct bearing on health care utilisation in issues in the countries of origin.
- *Regulatory policies* which promote rather than constrain the ability of traditional health services to continue contributing to the health care needs of the majority of the population of most countries, particularly the rural poor.
- *Training policies* which strengthen training within the epistemological framework of traditional health systems rather than training policies designed to promote re-training in modern medicine for traditional health care practitioners.
- *Biodiversity policy* which is designed to conserve medicinal plants and to generate appropriate harvesting and cultivation strategies. Biodiversity policies as they pertain to medicinal plants should recognise the widespread reliance of rural populations on medicinal plants for their basic health care. National and international herbal medicine manufacturers should be encouraged to assume responsibility for regenerating the sources of the plants which provide their source of *materia medica* and income.
- *Cultural and intellectual property rights* are perceived differently by Western and non-Western societies. Clear statements of the legal rights of individuals

and communities who participate in traditional medical research should be enunciated in the context of international accords designed to protect the rights of customary knowledge holders.

- *Inter-cultural exchange* of health care strategies and traditional medical knowledge. The experience of countries and communities which have taken steps to strengthen their indigenous health traditions can serve as a guide to others intending to undertake similar activities. International cultural exchanges, akin to international scientific exchange in biomedicine, should be supported by foundations and development agencies involved in health sector development.
- *Scientific, governmental and public education* on traditional health care. Policy makers, biomedical practitioners and scientists in both developing and industrialised nations need to be exposed to research and clinical experience in traditional and alternative medicine if genuine pluralism in health care is to be achieved in the context of international health policy reform.
- *Funding* is the *sine qua non* of the above recommendations. The current situation is one in which approximately 100 per cent of the health resources invested in and by developing countries goes to support services utilised by approximately 20 per cent of the population. Equitable distribution would require that a significant reallocation or increase in resource allocation to the traditional health sector be undertaken in order to adequately address the needs of this sector in servicing the health care needs of the majority of the population.

References

Akerele, O., Heywood, V. and Synge, H. (eds), 1991, *Conservation of Medical Plants*, Cambridge University Press, Cambridge.

Argueta, A., 1995, in *Indigenous Peoples and Development*, Davis, S. (ed.), World Bank, Washington DC.

Aryal, M., 1993, Diverted wealth: the trade in Himalayan herbs, HIMAL, Kathmandu, Jan/Feb p.10.

Berger, R., 1995, Traditional healers, *The Lancet*, 345, 25 March, p. 796.

Bishaw, M., 1991, Promoting traditional medicine in Ethiopia: a brief historical review of government policy, *Social Science and Medicine*, 33, 2, pp. 193–200.

Bodeker, G., 1994, Traditional health knowledge and public policy, *Nature and Resources*, 30, 2, Unesco, Paris.

Bodeker, G., 1998, Regulation vs Promotion in natural medicine: the impact of Northern policies and the need for a comprehensive policy development strategy. Herbal Medicine Regulation, in: *Regulating Herbal Medicine*, Shuman *et al.* (Eds), University of San Diego Press, 1998.

Bodeker, G., Willcox, M., Flohr, C. and Ong, C.-K., 1998, The therapeutic potential of plants used traditionally as antimalarials: new directions for research, in: *Health in the Commonwealth*, Commonwealth Secretariat, Kensington Publications Ltd, London.

Choe Won Sok, 1995, Pyonyang, Korea: Country Report, *WHO Symposium on the Utilisation of Medicinal Plants*, University of Pennsylvania Press, Pittsburgh, PA.

Cunningham, A.B., 1993, *African Medicinal Plants: Setting priorities at the interface between conservation and primary health care*, People and Plants Working Paper, UNESCO,

Division of Ecological Sciences, Paris.

Cunningham, A.B. and Mbenkum, F.T., 1993, *Sustainability of harvesting prunus africana bark in Cameroon: A medical plant in international trade*, People and Plants Working Papers, Unesco, Paris.

Davis, W., 1995, Book Reviews, *Herbal Gram*, American Botanical Council, 33, p. 66.

Duke, J., 1995, Plants as Medicines, *Journal of Alternative and Complementary Medicine*, 1, 1, January, p. 12.

Dutfield, G., 1997, Between a rock and hard place, in: FOA, *Medicinal Plants for Forest Conservation and Health Care*, Bodeker, G., Baht, K.K.S., Burley, J. and Vantomme, P. (Eds) UN Food and Agriculture Organisation, Rome.

Eisenberg, D.M., Davis, R.B., Ettner, S.L., Appel, S., Wilkey, S., Van Rompay, M. and Kessler, R.C., 1998, Trends in alternative medicine use in the United States, 1990–1997: results of a follow-up national survey, *JAMA*, 280: 18, 1569–1575.

FAO, 1997, *Medicinal Plants for Forest Conservation and Health Care*, Bodeker, G., Bhat, K.K.S., Burley, J. and Vantomme, P. (Eds), UN Food and Agriculture Organisation, Rome.

Farnsworth, N. and Soejarto, D.D., 1985, Potential consequences of plant extinction in the United States on the current and future availability of prescription drugs, *Economic Botany*, 39: 231–240.

Fellows, L.E., 1991, *Pharmaceuticals from traditional medicine plants and others: future prospects*. A paper presented at the symposium New drugs from natural sources, I.B.C. Technical Services Ltd., 13–14 June, Royal Botanical Gardens, London.

Flaherty, M., 1993, in: Pan American Health Organisation (PAHO) and Canadian Society for International Health, *Proceedings of Conference on Indigenous Peoples and Health*, Winnipeg, Canada.

Fugh-Berman, A., 1993, The Case for Natural Medicine, *The Nation*, September, 6/13.

Hall, R.L., 1986, Alcohol treatment in American Indian populations: An indigenous treatment modality compared with traditional approaches, *Annals of the New York Academy of Science*, 472: 168–178.

He, S.A., 1997, Country Report, *WHO Symposium on the Utilisation of Medicinal Plants*, University of Pennsylvania Press, Philadelphia.

Joyce, C., 1994, *Earthly Goods: Medicine Hunting in the Rainforest*, Little, Brown/Time Warner Books, New York.

Kabatesi, D., 1998, Use of traditional treatments for AIDS-associated diseases in resource-constrained settings, in: *Health in the Commonwealth*, Commonwealth Secretariat, Kensington Publications Ltd, London.

Kleinman, A., 1983, *Patients and Healers in the Context of Cultures*, University of California Press, Berkeley.

Kogozi, J., 1994, Herbalists open hospital, *New Vision*, Kampala, Uganda, 4 February, p.14.

Lange, D., 1998, *Europe's Medicinal and Aromatic Plants: Their Use, Trade and Conservation*, TRAFFIC International, Cambridge.

Le, T.T. and Bodeker, G., 1997, Traditional medicine and the infrastructure for burn care in Vietnam, *Tropical Doctor* 27 (Supplement 1), 39–42.

TraditionalLewington, A.A., 1993, *A Review of the Importation of Medicinal Plants and Plant Extracts into Europe*, TRAFFIC, Cambridge, UK.

Lewith, G. and Aldridge, D. (eds), 1993, *Clinical Research Methodology for Complementary Therapies*, Hodder & Stoughton, Kent, UK.

McNeely, J.A., 1993, People and protected areas: partners in prosperity, in: Kemf, E., *The*

Law of the Mother: Protecting Indigenous Peoples in Protected Areas, Sierra Club Books, San Francisco.

Mubiru, N., 1994, Ministry of Health, Uganda, Paper presented at *Gifts of Health conference, Mbarara, Uganda* (unpublished).

Nichter, M., 1978, Patterns of curative resort and their significance for health planning in South Asia, *Medical Anthropology*, 29–58.

Norbeck, E. and Lock, M., 1987, *Health, Illness and Medical Care in Japan*, University of Hawaii Press, Honolulu.

Parry, E., 1997, The scope and limits of traditional care, *Tropical Doctor*, 27 (supplement 1), 2–3.

Pelletier, K.A., 1991, Review and analysis of the health and cost-effective outcome studies of comprehensive health promotion and disease prevention programmes, *American Journal of Health Promotion*, 5, 4, March/April.

Peters, C.M., 1994, *Guidelines for the Sustainable Harvest of Non-timber Plant Resources in Tropical Moist Forest: An Ecological Primer*, Biodiversity Support Program, Washington, DC.

Principe, P.P., 1991, Valuing diversity of medicinal plants, in, Akerele, O., Heywood, V. and Synge, H. (eds), *Conservation of Medicinal Plants*, Cambridge University Press, Cambridge.

Reid, W. *et al.*, 1993, *Biodiversity Prospecting*, World Resources Institute, Washington DC.

Roy Chaudhury, R., 1993, *Herbal Medicine*, WHO Regional Office, Delhi.

Rural Advancement Foundation International (RAFI), 1993, *Endod: A Case Study of the Use of African Indigenous Knowledge to Address Global Health and Environmental Problems*, RAFI Communiqué.

Seeds of Survival, 1993, African Diversity, *Seeds of Survival*, 7 October.

Sotomayor Castellon, U., 1992, *Report of the Fundación Centro Nacional de Medicina Popular Tradicional*, Dr Alejandro Davila Bolanos Foundation Publication, Nicaragua.

Uganda Government, 1993, *National Drug Policy and Authority Statute*, No. 13, Part II, 3(1) (g).

UNDP, 1994, *Conserving Indigenous Knowledge: two systems of innovation*, United Nations Development Program, New York.

Wilson, A., 1993, Sacred Forests and the Elders, in: Kemf, E., *The Law of the Mother: Protecting Indigenous Peoples in Protected Areas*, Sierra Club Books, San Francisco.

Wootton, J., 1995, Education and Information Resource Development, *Journal of Alternative and Complementary Medicine*, 1, 2, 197–204.

WHO, 1978, *Traditional Medicine*, World Health Organisation, Geneva.

WHO, 1990, *Report of the WHO Consultation on AIDS and Traditional Medicine: Clinical Evaluation of Traditional Medicines and Natural Products*, Geneva, WHO/TRM/TRM/90.2.

Zolla C., 1993, The Cuetzalan Experience, Pan American Health Organisation (PAHO) and Canadian Society for International Health, *Proceedings of Conference on Indigenous Peoples and Health*, Winnipeg, Canada, pp. 1–64.

Author's Note

This chapter is reproduced with permission of the publisher and editors of the *Journal of Alternative and Complementary Medicine*, Vol. 1, 3, 1995. The material in the chapter also draws, with permission, on an earlier publication in Unesco's journal *Nature and Resources*, Bodeker, 1994.

11 Health of rural and urban communities in developing countries

A case study in Indonesia

Shosuke Suzuki

Introduction

It is necessary for the developing countries to concern themselves with population control, to enhance opportunities for universal education, especially for women, and to focus on environmental conservation for ecological sustainability.

Two villages in West Java and one in Central Java were studied intensively in the first half of the 1980s and were followed up until 1995. The villagers cultivate rice twice a year with little fertiliser or pesticide application. They make use of home gardens to cultivate fresh vegetables and fruit. They keep chickens, ducks and fish in ponds and rice fields in order to get animal protein.

Domestic waste including human excreta is consumed in the ponds by fish and used as manure for rice cultivation. Such organic recycling and mixed cultivation has made the life of these villagers a sustainable one for more than 2,000 years. They use well or spring water for drinking and cooking, and stream water for irrigation and washing. More than half of the villagers are infested with parasites, particularly ascaris and hookworm. This indicates incomplete digestion of human excreta in the pond.

The potential to increase productivity in the agricultural sector is very limited since no more arable land is available and hence there is no further opportunity for an increase in rice productivity. Thus the younger generation from such rural areas has been displaced to large cities, to outer islands other than Java, or to Saudi Arabia. More than 95 per cent of the villagers are Moslem.

The increasing population of the capital city, Jakarta, has made the urban environment worse. Water pollution, and air pollution from high traffic density, are major problems. The health impact of atmospheric lead has been measured. Drivers of small buses had twice the level of atmospheric lead in their blood and urine. Their blood enzyme levels of delta-aminolevulinic acid dehydrise were depressed. Roadside lead exposure for eight hours a day affected such exposed people at the cellular biochemical level.

A clean water campaign in Jakarta between 1992 and 1994 had limited success. Air pollution is capable of improvement by the application of technology and the use of legal restrictions on emission. Such actions are essential because Indonesia is developing at the rate of 6–9 per cent annually.

Almost all of the islands in the Pacific Ocean are currently undergoing rapid development. This is also true for Sumatra and New Guinea. Island populations have increased rapidly. Rainforest, fauna, and natural resources, have been consumed and exploited. Factories, roads, air and sea ports have been built using the 'development assistance' of some developed countries. This has caused further degradation of the natural environment.

Culture has been lost by the introduction of television, Western music and Western education. How and why are these environmental problems being caused in developing countries? How is sustainable development possible in these countries or regions? These issues are presented here, based on the author's study experiences in rural villages as well as in Jakarta.

An overview of the issues in developing countries

Some people say that developing countries may not necessarily want to be developed, and it is only the view of developed countries that they do. This may be valid only in special settings. If a national or an ethnic group did not develop, it would be far behind other developing countries, and often could be invaded.

Human groups on earth differ in habitat and culture. Their exchanges of information, food, goods, people, and technology have helped to diminish differences between countries. War has also led to decreasing such differences, while, at the same time, new classes and conflicts emerge. French social anthropologist Lévi-Strauss recognised that human history has always worked towards making the potential energy of the difference smaller.

South Korea, Taiwan, Hong Kong and Singapore are now considered to be developed countries. Thailand, Brazil and India are among medium developed countries. China, Indonesia, Malaysia and other Asian countries experienced rapid economic growth from 1985 until 1997.

After the Neolithic revolution large states were born, where many people lived sedentary lives on fertile lands. Hunter-gatherers could not construct large states because they had to follow their food supplies. Hence, the size of the group was, at most, dozens of people and had no resources to support people who did not engage in productive work.

Following the creation of modern nation states in Europe, these were expanded throughout the world by force through the formation of colonies. Political borders were drawn between these colonies all over the world. Nomadic groups faced extinction as a consequence of the imposition of such political borders. Every land in the world belonged to a colonial power.

Thus most current developing countries have been the colonies of host countries for a few centuries, and have been influenced by the colonising state in language, industry, organisation and human relations.

For example, in Java, some of the senior members of higher society can speak Dutch fluently. Old rubber tree forests are seen everywhere in Java, and there are tea plantations in the highlands. They are the remnants of forced cultivation from

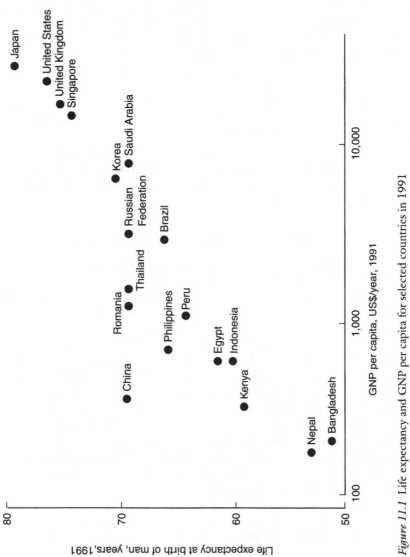

Figure 11.1 Life expectancy and GNP per capita for selected countries in 1991
Note
The horizontal scale is logarithmic

colonial times. The Indonesian language has been influenced by Dutch, and the Malaysian language by English.

Imposed state borders made survival for some nomadic groups difficult, since such groups lived by seasonal migration. Most of the ethnic groups in Africa and North and South America disintegrated and vanished – an annihilation.

The level of the development of a state is indicated by its GNP. GNP per capita of the poorest country is less than US$420. The world population of the poorest is 1.1 billion. GNP is not an indicator of living standard but of economic development. Figure 11.1 shows some of the world's poorest countries. Middle income countries of US$421 to $3,000 per capita GNP are also shown. The vertical axis shows average life expectancy at birth. The horizontal axis represents GNP per capita on a logarithmic scale. Each increase in life expectancy of one year requires a doubling of GNP per capita. China appears an exception to this observation. This may be caused by an under-registration or under-reporting of infant deaths, or an increased level of maternal and child health brought about by the wide availability of school education in China.

Population of developing countries and their environment

The population of the world was increasing at a rate of 1.7 per cent by 1994, although its maximum rate of increase was estimated to be 2.1 per cent between 1965 and 1971. The world population will be 6.2 billion by the year 2000. The human population continues to explode in Africa and South America.

Developed countries may experience slight population increases. Decreases are possible given that their total fertility rates (TFRs) have been below 2.0 for a few decades. The TFR of Japan was 1.46 in 1993. TFRs of Korea, Singapore, Taiwan, Hong Kong, Thailand and China are also less than 2.0. Other Asian countries have higher TFRs. Even when a country has a TFR of 2.0 or less, the population may still continue to rise. This is due to the inertia of population growth.

Family planning is becoming more popular and accepted. However, the world's population will continue to increase, at least until the middle of the twenty-first century, with a peak of 8 to 10 billion. Human beings will experience various difficulties in the future including scarcity of resources, pollution of the environment, health problems, religious and ethnic conflicts.

A great gulf between the poorest and richest countries may cause serious problems in the future. The population increase in developing countries will depress social development. As these people sell natural resources their environment will become degraded, their social development will be delayed, and the socioeconomic gap will be widened between these developing and developed countries. Migrants or refugees are a further burden in developing countries. Such population increase is the starting point of serious problems.

What is the cause of the population explosion?

Traditional societies had high birth rates and high death rates before the introduction of modern technology to those societies. The colonisers or the suppliers of development aid brought modern technology for environmental sanitation, safe

Table 11.1 Vicious and healthy cycles of 'education–population–resource production'

Vicious cycle of negative feedback	*Healthy cycle of positive feedback*
Many children	Education
Population increase	↓
↓	Family planning
Food shortage	↓
↓	Small family
Malnutrition	↓
↓	Basic human
Infection	needs satisfied
↓	↓
High infant mortality	Take-off for
Large families	industrialisation
↓	and modernisation
Population increase	

Note
Education is the key to dismantling the vicious cycle.

drinking water, sewage treatment, pesticides, antibiotics, vaccination, health education, and so on. Everything but family planning. Those technologies caused a decrease in infant mortality, although the birth rate changed little.

People in developing countries suffer from having had many children in a household. They have no more arable land to cultivate. Children must leave villages for large cities. The author once heard a middle-aged Indonesian farmer complaining he wished he was knowledgable about family planning before he and his sick wife had six babies. His wife died of tuberculosis, leaving him with six small children.

The author lived intermittently in West Java from 1980 to 1985. More than half of the tenant farmers there have no farm land. A few rich villagers own more than half the land, where wet rice cultivation is their major source of food and income. The population increase has caused them to cultivate almost to the ridge line of the hills.

The preserved forest owned by the central government is often cut down by villagers to get firewood for their own use or sale. The forest land is often eroded from heavy rainfall. Soil and sand silt up their wet rice fields and streams. Agricultural wage-labourers or tenant farmers cannot feed their excess children, nor allow them to attend primary school. Babies are killed by diarrhoea or pneumonitis. The infant mortality rate in 1980 was estimated to be 180 per thousand live births.

Following the building of a new primary school in their rural community, most villagers, although facing much hardship, have encouraged their children to attend. After graduation children go to town to assist their relatives who work there selling food, medicines and goods. The increased population causes yet other problems in the cities. This is a vicious cycle of negative feedback as depicted in Table 11.1 (left).

How can this vicious cycle be transformed into a healthy cycle with a positive feedback as in Table 11.1 (right-hand side)?

The first priority should be support for the education of women, integrated with community development programmes. The World Bank started a population policy in the 1980s recognising that there can be 'no community development without family planning'. Investment for community development proceeded in Indonesia during the 1980s, supporting community organisation, the establishment of women's associations, and the creation of primary health care centres.

Family planning calls for the constructive participation of husbands, the integration of family planning into the community's annual programmes, the provision of a variety of contraceptive goods from the government, and obtaining the approval of the religion sector from grass roots right to the national level.

Integrated and repeated approaches are necessary for successful family planning. In Indonesia, the acceptance rate of family planning was less than 50 per cent in 1980. This rose to almost 90 per cent in 1990, although there remain large differences according to region and social class. As a result, the population increase rate of 2.3 per cent for Indonesia in 1980 decreased to 1.9 per cent in 1990. Family planning is securing increasing acceptance in developing countries world wide.

China, with a population of 1.2 billion in 1994, has imposed the one-child policy. This policy is contributing to social development. A reduction in the number of children allowed the Chinese people to buy electrical goods for use in the home. Factories to manufacture the goods have been established. They export the products. They employ more Chinese people, who thus become wealthier and better educated. They are trying to promote their social development.

On the other hand, some side effects of the rapid development have become evident. Rapid industrialisation has caused many industrial accidents and diseases throughout China, especially in village factories. Coal burning and inefficient incineration have resulted in serious air pollution. Lowland development and deforestation have caused serious land erosion, catastrophic floods and the massive extinction of wild life. These current method of development can never lead to sustainable development.

Health and health outcomes in rural Javanese villages

Ecology and villagers on the island of Java

Java is a volcanic island with an area of around 120,000 square kilometres. It is situated in the archipelago of the Republic of Indonesia which includes also the islands of Sumatra, Kalimantan, Sulawesi, Maluku, Nusa Tenggara, and Irian Jaya as shown in Figure 11.2. Java has a tropical climate, being located between 6 and 8 degrees south of the Equator.

In the lowland areas of Java, the maximum daily temperature is 30–33 degrees Celsius and the minimum is 23–27 degrees Celsius all year round. Temperatures below 20 degrees Celsius are not experienced. The Javanese seem less able to

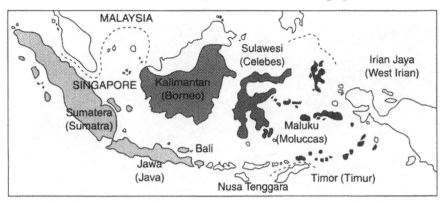

Figure 11.2 Map of Indonesia, showing major islands and regions

adapt to changes in temperature than those living in temperate zones. In this tropical zone, there is no pronounced seasonal wind.

The island of Java is mountainous. Two-thirds of the island is highland.
Traditionally people have lived there because the climate is cooler and water management is easier. The colonial Dutch loved and enjoyed Bandung, located at an
altitude of 800 metres. Bandung is cool, fertile, and full of fruit and flowers. It is
in the centre of the Priangan Highlands, where it is the homeland of those West
Javanese people called Sundanese. The Priangan Highlands have many volcanic
mountains higher than 2,000 metres. These volcanoes emit ash, clay, marl and
lava. These are full of minerals and make the Javanese land very fertile.

Farmers cultivate rice in terraced fields as shown in Figure 11.3. They cultivate
various traditional rice varieties in a rotational seasonal scheme. This practice
decreases the risk of pests. Traditionally they do not use chemical fertilisers. They
harvest only the grain at the top of the rice plant, and the other parts of the rice
plant are buried and composted. Recycling of organic materials and close sym-
biosis with the rice plant is a good example of how sustainable yields can be
obtained.

Java is very fertile. Its original vegetation was tropical rainforest. A new migrant
group from continental Asia came to Java about 2,000 years ago bringing a rice
cultivation technology. Only in Java and Bali was rice cultivation successfully
developed. This is because the soil there was very fertile. Borneo island was also
covered with thick tropical rainforest, but the soil is old and sandy, weathered and
completely leached. Borneo has no volcanoes able to supply fresh minerals.
Organic matter is not in the soil but is held in the vegetation. Nutrients produced
by the detritus are rapidly absorbed by the vegetation.

In the northern coniferous forests, the leaves and litter have accumulated on
and in the topsoil for many years. When these forests are cut, organic substances
are lost for ever and the poor sandy soil is left which is not suitable for rice cultivation.

Figure 11.3 Rice fields in West Java

Table 11.2 Population, area, and population density by region in Indonesia

Island	Population × 10³	Area × 10³	Population density /km²
Sumatra	36,472	473	77
Java	107,527	132	814
Nusatenggara	10,162	88	115
Kalimantan	9,096	539	17
Sulawesi	12,509	189	66
Maruku and Irian Jaya	3,483	496	7
Total	179,379	1,919	93

Source: Central Bureau of Statistics, Indonesia, *Statistics of Indonesia 1991*.
Note
Population density is dependent on soil fertility

In Indonesia and Malaysia tropical rainforest is sometimes subjected to slash and burn agriculture. If limited in area, this practice is not completely destructive of vegetation. Cyclical use is possible, and can result in a sustainable yield. However, modern incineration methods now applied in Brazil are not sustainable.

Tropical rainforest could be used most effectively if fruit and nuts were harvested in a sustainable manner. Timber thinning is also a sustainable use of forests. In the 1980s, the Indonesian government commenced reafforestation in Kalimantan (Borneo) with financial aid and technology from Japan. As this is the first trial, the successful regeneration of tropical rainforest in Kalimantan is not guaranteed.

The soil of Java is so fertile that rice cultivation is possible even after tropical rainforest clearance. Additional nutrients are supplied by streams from high mountain forests. Human excreta is recycled in ponds, the effluent nutrients being used again by rice plants.

Java has the highest population density in Indonesia with 814 people per square kilometre, while that of Kalimantan is 17 per square kilometre (Table 11.2). The people have developed sophisticated rice cultivation practices, underpinned by a complex culture reflected in their social organisation, kinship systems, customs and mythologies.

The water cycle in a rural village

The original vegetation in Java was tropical rainforest, and the island has two seasons, wet and dry. It sometimes suffers from drought in the dry season. Usually there is ample rain and water. In the Indonesian language the natural habitat is *tanah air* meaning land water (fatherland). The people believe land and water are integrated concepts and cannot be separated from farmers, animals and vegetation. Water is used not only for rice cultivation, cooking, drinking, and washing, but also for *mandi* or Islamic bathing.

The map shown as Figure 11.4 illustrates the water cycle in a village in mountainous West Java. Salamungkal village is located near the forest reserve of Mount Salasih, with the peak of 1,944 metres. The village is built on a ridge with a south to north slope of 6 to 9 degrees. Water for the village has been diverted from a stream at a higher altitude. This system of water supply guarantees a safe and sunny habitat for farmers. Water is used for irrigation and collected in ponds. The people use spring water for drinking and cooking. Girls and women do the carrying. In the tropics mountain ridges are a comfortable and hygienic habitat wherever ample water is available.

Mapping is recommended as a priority task for participant observation field workers. They become acquainted with every place and every villager and become accepted as a person capable of drawing precise maps of the habitat. A team member, Mr. Igarashi, constructed the map in Figure 11.4 over a period of several weeks.

In the centre of the map, a main street runs along the mountain ridge. A stream runs beside the main street. Squares represent the farmers' homes, and circles represent ponds. There are various sized ponds near the houses. Other symbols represent orchards, trees, ornamental vegetation, vegetable gardens, and hedges around the houses. The gravitational flow of the water is from south to north.

'S' on the map symbolises the spring. It provides germ-free, non-polluted drinking water. Spring water is led along a bamboo pipe for about 50 metres to a watering hole at a spot marked on the map as 27. Women go to that spot to draw water each morning and evening (Figure 11.5). The spring water in the tank is used by about 50 households.

Small circles mark washing places on ponds (Figure 11.4). There they wash their bodies, clothes, vegetables, dishes, and so on. Triangles mark pond latrines.

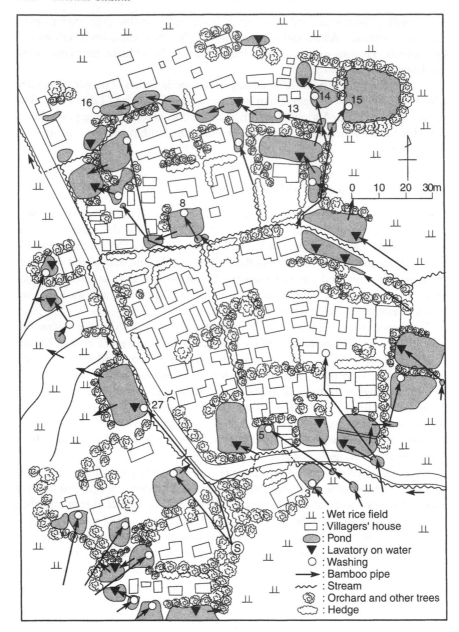

Figure 11.4 Map of Salamungkal, a village in West Java

Those ponds are never a source of washing water as seen in Figure 11.4. However, polluted pond water does flow down to the rice fields.

Carp are kept in latrine ponds and digest human excreta. Effluent from ponds contains large amounts of fertiliser together with parasite eggs. The ponds are

Figure 11.5 A washing place by a pond
Note
Water flows from the rice fields. Clothes, dishes and pots are washed here

almost 1 metre deep. The total area of ponds amounts to 4,202 square metres, giving 44.2 square metres per household. This space is larger than the average land area occupied by a household (35 square metres). Households average 4.5 persons.

Drinking water

The villagers use water from a spring for drinking and cooking. This spring water is pure and free from human excreta pollution. All water from wells in other rural communities is usually polluted.

The spring site marked 'S' in Figure 11.4 was used by about two-thirds of the people, about 50 households. The remainder of the village used another spring which was under a small cliff. It produced only 10 litres per minute. The distance from a house to spring water and the water volume from a spring determine which spring particular people use.

Indonesians do not directly drink river water, tap water, well water, or spring water. They only drink boiled water or weak tea. This is germ free and is served to guests in the village. Villagers usually dry cups and dishes in sunny places which sterilises them. In cities iced water or cold home-made juice is often served. This frequently causes diarrhoea. In cities ice often contains bacteria, and cups and dishes are often washed with polluted water and not disinfected by sunshine. Diarrhoea is more prevalent in the cities than in the villages.

When suffering from diarrhoea, the priorities for successful treatment are rest, oral rehydration with replacement of salts, and as a last choice antibiotics are

Figure 11.6 A typical well used in villages in West Java
Note
Water is drawn in a bucket with a rope. Several nearby households use the well.

sometimes used. In every rural community, one person keeps powdered oral rehydration medicines which are usually very cheap. A villager suffering from diarrhoea visits the medicine keeper in the community. They dissolve the powder in clean water. This is used for all age groups, especially for young children. Rehydration has saved the lives of many young children in Indonesia. Two concepts have dramatically decreased infant mortality there: health education on the hygienic preparation of food and drink, and the system of rehydration powder distribution and use in every rural village.

In large cities in Indonesia, tap water is often polluted as pipes leak a significant amount of water which becomes polluted by sewage when water pressure is low. Small restaurants and food vendors at the roadside often sell contaminated food. It is difficult for city residents to avoid diarrhoea.

Well water

Villages in lowland Indonesia usually have no springs. Instead they dig wells, usually shallow ones, with no roof or well cover (see Figure 11.6). The area surrounding the well is unlikely to be concreted, nor are there properly lined drainage ditches. These wells are often used by several nearby households, but are often empty in the dry season.

Deep wells supply higher quality water, but villagers have neither the technology nor the money to build them. The author saw a deep well in an elementary school in Central Java financed by development aid, which supplied good quality water.

Well water is used for cooking, washing, bathing and sometimes for irrigation. Villagers make a bathing place next to the well. Often there are two or three such places since so many people use them. Each has a floor made from concrete and is surrounded by walls and a door made from bamboo or concrete. Muslims bathe up to three times a day before every prayer time.

The quality of well water is better than washing water taken from rice fields, and worse than spring water. In Salamungkal, they use spring water for drinking and cooking and stream water for washing. Villagers in the lowlands dig wells and use this water for drinking and washing. Neither highland nor lowland villagers drink water without first boiling it.

Water for washing

Water for washing is very important for villagers, since they are Muslims. Custom dictates that they wash their bodies, especially their eyes, nose, mouth, and private parts, five times a day. Women are dexterous in washing their bodies without revealing them to the public.

In Salamungkal the people use water from rice fields, channelled to a small pool where it is used for washing in cages made of bamboo (Figure 11.7). Villagers wash rice, vegetables, dishes, and clothes in the cage with this running water. Before prayer, villagers sometimes line up in front of the washing place.

Even though the quality is poor, water flows continually even in the dry season. Table 11.3 demonstrates that turbidity, precipitates and colour are a problem. The water contains significant amounts of ammonia when it comes from a pond with a latrine. Such water is polluted by human excreta. The water carries bacteria and parasitic eggs.

Poor villagers use polluted water. Rich villagers use cleaner water since they take clean water from a stream, using long distance bamboo pipes. Poor people often use cleaner washing places owned by richer villagers.

Water for irrigation

Wet rice fields surround a cluster of community houses. The fields are irrigated from the stream which flows from the mountain ridge. Water from higher rice fields runs down to the lower, terraced, wet rice fields (Figure 11.3). Water also flows down to ponds which villagers use for latrines and fish keeping. This pond water flows down to the rice fields. The water is effectively recycled. Water from the mountain never dries up, and is always available even in the dry season, while rice fields on the plains suffer from shortages of irrigation water.

The mountain villagers cultivate traditional rice varieties. They can cultivate rice at any season of the year, because water is available all year round. The temperature ranges between 20 degrees and 30 degrees Celsius which is suitable for rice plants. All stages of rice plant growth can be observed at any time. This decreases the risk of pest damage. A variety of foods are continuously available.

Villagers have dry fields on the steeper slopes. They cultivate many kind of

Table 11.3 Quality of drinking water and washing water used in Salamungkal

Sampling spot, no.	Colour	Turbidity	Precipitates	Ammonia	Total score	Temperature, Celsius	pH	Source of water, and comments
Water used for drinking and cooking								
26	5	5	5	10	25	23.9	7.2	Spring water, piped 30m
27	5	5	4	4	18	23.0	7.7	Spring water, piped 70m
28	5	5	5	10	25	24.9	6.2	Spring water, unpiped
Water used for washing								
3	4	5	5	10	24	21.2	7.5	Rice field
5	5	5	5	10	25	21.5	7.5	Rice field
8	4	4	3	10	21	22.5	7.5	Stream
11	5	4	4	10	23	22.9	7.5	Stream
13	4	3	3	2	12	22.0	7.5	Stream and pond with waste
14	4	3	3	2	12	22.0	7.4	Stream and pond with waste
15	5	4	3	8	20	22.9	7.5	Stream only
16	4	4	3	2	13	23.0	7.5	Five ponds with waste

Notes

The sampling spot numbers correspond with those on the map (Figure 11.6).

Colour: 5 is colourless, 4 is light yellow, 3 is light brown, 2 is brown and 1 is dark brown.

Turbidity: A 9-point letter may be read through the water column of 140 mm or more (5), 101–140 mm (4), 51–100 mm (3), 20–50 mm (2), and less than 10 mm (1).

Precipitates: No ppt or very little (5), detectable (4), a little (3), much (2), and very much (1).

Ammoniacal nitrogen by Nestler method: less than 0.2 ppm (10), 0.2 ppm (8), 0.2–0.5 ppm (6), 0.5 ppm or more (2).

Ammoniacal nitrogen was detected most in water from latrine ponds. Spring water nos 26 and 28 are potable with nos *E. coli*; no. 27 is not potable, containing much ammoniacal nitrogen and *E. coli.*

Figure 11.7 A farmer and his daughter-in-law fish, with a net, from a latrine pond

plants: tobacco, onions, talus, red pepper, young orange trees, papayas and banana trees. This is carried on as mixed cultivation in small steep fields, and continuously supplies various kind of foods for the villagers. This practice creates sustainable yields.

Women villagers harvest only the top or grain part of the rice plant. The other parts of the rice plant, the leaves, stalks and root, are left in the rice fields. When the next water cycle arrives, small fauna, insects, and water animals return to live on the rice plant. A flock of manila ducks, kept by a keeper-farmer, walk on the fields eating grasses and animal fodder. They lay eggs. The keeper sells the eggs and poultry meat to villagers. The ducks defecate in the rice fields. This is useful as a fertiliser. The decaying rice plants serve in part as compost. These plants are mixed with rice field soil by ploughing and become natural fertiliser. Water from the forest reserves also contains plant nutrients.

Ecological rotation works well in the lives of the villagers. Villagers subsist on this Javanese agricultural system without using artificial fertilisers or pesticides. They have maintained this lifestyle for more than 2,000 years. This is surely an example of 'sustainable yields'.

Fish ponds and water: a sewage treatment system

Making a pond next to their homes is a tradition of the Sunda people. A pond is used both for inland fishery and for sewage treatment (Figure 11.7). In the pond they keep carp or ikan mas, tilapia or ikan mujaer, and so on. When they have an unexpected and important guest, they serve a few fish from the pond, deep fried.

They cook and eat them on Idul Fitri, one of biggest festivals of Moslems. Faeces and kitchen garbage are digested by the fish and bacteria in the pond to yield various kinds of nutrients, especially phosphate, potassium, and nitrate, which flow to the rice fields and are used by rice plants. Thus the pond brings three advantages: sewage treatment, fish and fertiliser for the villagers.

Materials recycling is completed by human activity and the system has been sustained for the past 2,000 years. However, it is difficult to increase productivity with this system due to limited availability of land and labour intensive agriculture. This system cannot produce any surplus of income to use for education, improvement in nutrition, or for further modernisation.

For their subsistence, the Japanese people had also depended on rice, other grains, and starchy root crop cultivation until the end of the nineteenth century, when they started to modernise. In 1950 Japan still had more than half of its labour force engaged in the agricultural sector. Human waste of the city dwellers was removed and carried manually by farmers to their village, where it was used as fertiliser after being composted or anaerobically digested. Villagers presented city dwellers with harvested vegetables or rice as a token of appreciation for the human waste. Nutrient recycling was completed between village and town in this traditional system in Japan.

When the population in towns and cities increased and cheap chemical fertiliser became available, farmers did not need the human waste any more. This traditional system of recycling was abandoned. Farmers bought chemical fertilisers. The chemical fertilisers applied to the fields were easily washed away by rainfall to rivers and sea. The rivers and bays have suffered from eutrophication. Human waste is incompletely disposed of and discharged into the rivers and the sea. Resources for making chemical fertilisers are consumed. Pesticides are used, and chemical fertilisers and pesticides are added to the soil. The farmers need more cash for agribusiness. The agricultural soil in Japan has been depleted of much organic substance and humus in the past 40 years. This system is an open one, and will not guarantee a sustainable yield.

Water quality and endemic parasites

The quality of water for drinking and washing is as shown in Table 11.3. Three drinking water samples were taken from two springs and judged good for drinking. However, most water for washing contained a large amount of suspended solids, ammonia and *E. coli*, and had a high turbidity. Cleaner water for washing came from a stream which is far from residences. Water for washing came from latrine ponds. Because of incomplete waste treatment this water was polluted by human excreta. Water for washing certainly contains parasite eggs from whipworm, hookworm and ascaris (round worm). Treatment time from defecation to water release is too short for parasite eggs to be eliminated. Young hookworm penetrate farmers' hands and feet when cultivating and weeding in wet rice fields. Vegetables washed with this water were contaminated with ascaris eggs.

In Salamungkal, 50 per cent of the faeces of villagers tested positive for eggs of ascaris and 58 per cent were positive for hookworm. This is considered usual under these environmental conditions. By comparison, the faeces of villagers on the floodplain of Central Java contained 15 per cent and 16 per cent, respectively, of these parasites.

In this village, human waste was disposed of in a pit toilet with a constructed cover. It used a water system similar to a Western style flush toilet. The people of Central Java do not use a pond disposal system in their culture. They no longer re-use human waste as a fertiliser. They use synthetic chemical fertilisers and pesticides on high yield rice varieties developed by the International Rice Research Institute (IRRI) to increase rice yield.

This village previously suffered from flooding of the Bungawan Solo river. Flood water accumulated to a depth of several metres during the rainy season for a month each year. Rice was harvested only once a year. The government of Indonesia constructed a large dam, Uonogiri, on the upper part of the Bungawan Solo river. After construction of the Uonogiri dam, water flow was regulated and flooding eliminated.

To increase fertility, the soil required a greater amount of chemical fertiliser to maintain rice production. This increased rice productivity brought farmers increased income which promoted primary school education, family planning, community organisation, kerosene for kitchen ranges, deep well boring, better diets, and improved lifestyle including the human waste disposal system.

The people of Central Java enjoyed a healthier lifestyle than the Salamungkal mountain villagers, who were unable to increase their income because of limited land and increasing population. As well as this, powerful Islamic traditions continue to govern the daily life of the Salamungkal villagers.

Health and health outcomes in urban Jakarta: air pollution

Ecology and urban life in Jakarta

Rural problems and urban problems have many characteristics in common. Further development of rural Javanese villages is severely limited. There is no more arable land available. There is an ever increasing birth rate. Only small improvements in agricultural production are possible, in spite of dam construction, the introduction of new rice varieties, and the use of chemical fertilisers and pesticides. This system of labour intensive Javanese agriculture has been at a maximum steady state for more than a century.

The government of Indonesia has promoted an out-migration policy for more than 30 years. Many young couples from villages migrated from their native place to Sumatra and Kalimantan, attracted by the offer of 2 hectares of arable land, a house to live in, and the basic necessities of life. The number of out-migrants was small compared with population growth in Java, and some of them returned to their native villages, unsuccessful, as the land was too infertile for their subsistence.

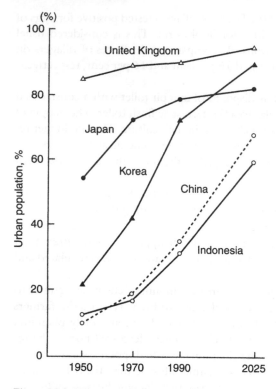

Figure 11.8 Urban population as a percentage of the total population of selected
countries

There are push factors in rural communities, and pull factors in urban settings. Large scale migration occurs in every developing country. Relatives of an initial migrant come to the city and assist in the small shop of that migrant or are employed by his pedicab group. A successful migrant employs workers from his native community. Migrants to urban areas increase in number, and are better able to survive in growing cities than in their native village. Some may become successful. Youngsters in rural villages have no prospects of a successful life as more than half of the household heads have no land. Thus, the urban populations increase rapidly as shown in Figure 11.8.

Urban problems usually arise from rapid population growth and industrialisation. Government lags far behind in providing the necessary housing, a safe water supply, sewage treatment, primary school facilities, transport, safety and health policies. The slum areas and numbers of squatters increase. Air and water pollution becomes serious.

Here is a case study from urban Jakarta. Jakarta is located on the northern plain of West Java. It is situated 6 degrees south of the Equator, and 107 degrees east of the Greenwich meridian. The altitude of the city averages 10 metres. A few small rivers flow from north to south into Jakarta Bay.

Jakarta city has an area of 750 square kilometres, almost all of which has been urbanised. The city was founded in the early seventeenth century as a political base of Dutch government. The port, market, and Chinatown first developed in northern Jakarta near the present port. The centre of the city moved to the south as the sea prevented further development to the north. Jakarta is expanding along large roads on the outskirts forming new satellite towns: Depok, Bogor, Tangerang, and Bakasi cities.

The climate of Jakarta is tropical with 1,760 mm of mean annual rainfall. The original vegetation is tropical rainforest. Two seasons are distinguished: rainy and dry, though there is considerable rainfall even in the dry season. The average daily maximum temperature is between 31 degrees Celsius in September and 33 degrees Celsius in December. The average minimum temperature is between 22 degrees Celsius in December and 25 degrees Celsius in August. The wind is westerly from December to March, easterly from April to August, northerly from September to November. The wind velocity is tropically gentle or 0.9 to 1.1 metres per second on average.

Jakarta city was not settled before the Dutch period, as the climate was too hot and the marsh land too difficult a site for construction. People preferred to live in the Priagan Highland to the south of Jakarta, where the climate was comfortable, fruit was plentiful, and water supply was simpler to manage.

The author lived intermittently in and surveyed Jakarta from 1980 to 1995. The most pronounced changes taking place during that time were the expansion of the city to the suburbs, and the rapid increase in vehicle numbers. Pedicabs were prohibited in the city as they were said to impede traffic. Motorcycles, tricycles, automobiles, and various kinds of buses all increased.

Mass transport in Jakarta is by main roads only as there are few underground or overground railways. Jakarta has about five main roads of five lanes with a pavement named Jalan Utama Raya, lengthwise and crosswise. These form a network, with medium and small roads and with highways. People commute exclusively in motor vehicles, public and private buses, and motor cycles and cars. Peak-hour traffic jams grow longer and commuting times are increasing due to a greater distance from the place of residence to the workplace.

The traffic volume of the main road in 1980 was 4,450 passenger cars, 4,848 motorcycles and tricycles, 360 buses and trucks, 120 bicycles and 1,230 pedestrians per hour during the daytime on the average weekday, barring traffic jams. Most people use buses.

Air pollution in Jakarta

Over 30 months from 1978 to 1981, total suspended particles (TSP) measured in Kayu Manis, a residential area near downtown Jakarta, was 257 µg/m³. During the same period in Pulogadun, a newly developed industrial district, the figure was 106.7 µg/m³. This industrial area is east of Jakarta and had few polluting industries. These two air pollution monitoring stations, including maintenance systems, were established in Jakarta as one of the global environmental monitoring stations.

The city government of Jakarta also measured TSP during 1981. The measurements were as follows: Gulodok, the busiest shopping district, 448 μg/m^3; Bandungan, a mixed commercial and residential area, 459; PGM, office district, 216; Monas, a city park in the centre of Jakarta, 126; Ancol, a recreation district along Jakarta Bay, 129; and Halim, near the international airport in the south suburbs of Jakarta, 104. These figures indicate that heavy traffic areas are polluted by automobile exhaust.

Jakarta has few factories. The most heavily polluted districts were Gulodok and Bandungan where the largest shopping centres are sited along the busiest streets extending from the railway station. Large numbers of motorcycles gather and emit pollutants in small parking lots near these centres. The levels of the other pollutants were: carbon monoxide 71–111 ppm, lead 90 μg/m^3, and maximum oxidants 0.159 ppm.

Lead pollution and its health effects

In Indonesia only leaded petrol is used. Petrol and light oil are sold by a national monopoly named Pertamina. No lead free gasoline is available in Indonesia. Lead burned in engines is emitted as suspended particulates into the atmosphere. Lead content parallels TSP.

In 1985, the author surveyed 30 minibus drivers, and used 30 farmers, lived in the suburbs of Jakarta, as controls. The drivers worked more than 10 hours per day and were exposed directly to exhaust gas on the streets. The farmers breathed air containing less than 1 μg/m^3 of lead. The lead content in the atmosphere near minibuses was 2–5 μg/m^3. Theses drivers inhaled particles containing lead from the atmosphere. Lead becomes attached to the bronchioles and lung alveoli, and is absorbed into the blood.

Average blood lead level in the driver group was 18.5 μg/dl, and 9.0 in the control population. The driver group had twice the lead level of the controls. Lead in the blood is known to disrupt the enzyme named delta aminolevulinic acid, which is used in haemoglobin biosynthesis.

This enzyme activity was measured. It was reduced on average by 30 per cent more in the driver group than in the control group. However, haemoglobin content did not fall in the driver group. This shows that lead had not affected the haemoglobin content (Figure 11.9).

Between 1991 and 1993, an experiment was carried out to determine the effect of automobile gas on roadside visitors. Two Japanese were exposed on the roadside in Jakarta for 8 hours per day for 3 successive days. The effect was evident. Lead in urine increased after the experimental exposure, and delta-aminolevulinic acid in urine also increased (Figure 11.9).

Residents of Japan are the people in the world least exposed to lead. This might be a reason why the difference in exposure was the maximum between Japan and Indonesia. Leaded petrol increases the efficiency and power of gasoline engines. Japan adopted lead free petrol in the 1970s.

A newspaper article in 1967 disclosed the danger of lead poisoning in Tokyo

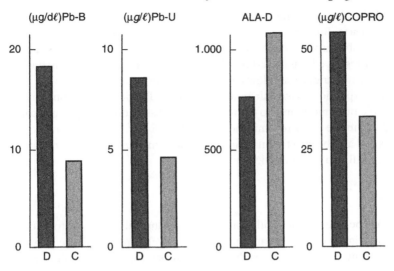

Figure 11.9 Health effect of lead in atmosphere polluted by automobile exhaust gas
Note
D = drivers; C = control group (farmers). The level of lead in the blood of drivers is twice that in the blood of farmers.

citizens. This was later found to have been a misinterpretation. A social movement for lead free petrol in Japan developed at this time, and was finally successful. The improvement of petrol by the use of the alternative additives and the improved design of engines resulted from a collaboration between the petroleum and automobile industries.

In 1990 after evaluating the report of the experiment on Japanese visitors, the Indonesian government decreased tetra-ethyl lead content in petrol by 50 per cent. This reduction had little effect on engine power. Lead content in the roadside air decreased by 30 per cent. The Indonesian government plans to introduce lead free petrol into the market, using alternative additives in place of tetra-ethyl lead. A lead free system will soon prevail in Jakarta.

Solving the problem of air pollution

To solve the air pollution problems, other pollutants need to be reduced. These include total suspended particles, carbon monoxide and nitrogen oxides. The number of private vehicles in the city must be decreased.

The municipal and national governments have attempted to decrease air pollution. Old engines produce more pollutants. Deficient filters also produce more pollutants. Inefficient engines emit incompletely burned gases and particles. An improvement in car engines and regular tests on vehicles are working to improve pollution emissions. Driver-only cars going into downtown Jakarta are heavily taxed or prohibited. Some drivers pick up a street child as a passenger, who then returns to the street after passing the checkpoint.

A kerbside bus lane is created each morning since buses are the major form of transport in Jakarta. Buses run continuously during the morning rush hour. There is no effective rail transport. Motor cycles weave between cars in traffic jams. These counter-measures have been undertaken by the government. The introduction of new high efficiency engines and low emission cars is in progress.

Rapid city expansion makes air pollution worse. Traffic jams become more serious. People need more time to commute. The construction of high rise apartment buildings in the city is recommended. The transfer of government offices to the suburbs will also be effective in reducing traffics jams. Restriction of migration from villages will also be required.

Conclusion

Indonesia has been growing rapidly since 1985. Industrialisation, motorisation, the spread of higher education, and mass media development – all are growing rapidly. The annual economic growth rate is 5–8 per cent. The middle class is increasing.

The government funded a large campaign, 'Clean water', in Jakarta in 1991–93. This was very successful in involving citizens and encouraging their participation in the solution of environmental problems.

The socio-economic gap between urban and rural life has become larger with the economic growth which has occurred since 1985. Appropriate development and sustainable, balanced change of village life will be the key to a healthy future for Indonesia.

Bibliography

Djunagsih, N., Hendrarto, Soemarwoto, O., Koyama, H., Hyodo, K. and Suzuki, S., 1988, Air pollution by lead and the health effect in Bandung City, in, Suzuki, S. (ed.), *Health Ecology in Indonesia*, Gyosei Co., Tokyo, pp. 203–212.

Hattori, T., Suzuki, S., Koyama, H., Ogawa, M., Aoki, S. and Igarashi, T., 1988, Differential growth of body height and weight of rural and urban children, in: Suzuki, S. (ed.), *Health Ecology in Indonesia*, Gyosei Co., Tokyo, pp. 31–40.

Kawada, T., Tri-Tugaswati, A., Kiryu, Y. and Suzuki, S., 1994, Relationship between mean lead levels in the atmosphere and in blood from data published since 1977, *Asia Pacific Journal of Public Health*, Singapore, 7(4): 233–235.

Kiryu, Y., Tri-Tugaswati, A., Suzuki, S. and Kawada, T., 1996, Health effects of atmospheric lead in roadside exposure in Jakarta, *Japanese Journal of Hygiene*.

Koyama, H., Moji, K. and Suzuki, S., 1988, Blood pressure, urinary sodium/potassium ratio and body mass index in rural and urban populations in West Java, *Human Biology*, 60: 263–272.

Suzuki, S., 1983, Conjunctivitis due to cultivation work observed among Indonesian peasants, *Journal of Human Ergology*, Tokyo, 12: 55–63.

Suzuki, S., 1987, Complex of environment, activity, and health in Indonesian Kampongs, in, *Health Ecology of Health and Survival in Asia and the South Pacific*, Suzuki, T., Ohtuka, R. (eds), University of Tokyo Press, Tokyo, pp. 149–164.

Suzuki, S. (ed.), 1988, *Health Ecology in Indonesia*, Gyosei Co., Tokyo.

Suzuki, S., 1990, Health effects of lead pollution due to automobile exhaust – Findings from field surveys in Japan and in Indonesia, *Journal of Human Ergology*, Tokyo, 19: 113–122.

Suzuki, S., Borden, R. J. and Hens, L., 1991, *Human Ecology – Coming of Age: An International Overview*, Proceedings of the Symposium Organised at the Occasion of the Vth International Congress of Ecology (INTECOL), Yokohama, Japan, August, 1990, Vrije Universiteit Brussel Press (VUB-Press), Brussels.

Suzuki, S. and Koyama, H., 1988, Health ecology of Sundanese and Javanese villagers in relation to social development and environmental factors, in, Suzuki, S. (ed.), *Health Ecology in Indonesia*, Gyosei Co., Tokyo, pp. 87–110.

Suzuki, S., Soemarwoto, O. and Igarashi, T. (eds), 1985, *Human Ecological Survey in Rural West Java in 1978 to 1982*, A Project Report, Nissan Science Foundation, Tokyo.

Tri-Tugaswati, A., Suzuki, S., Koyama, H. and Kawada, T., 1987, Health effects of air pollution due to automotive lead in Jakarta, *Asia-Pacific Journal of Public Health*, Singapore, 1: 23–27.

Tri-Tugaswati, A., Suzuki, S., Kiryu, Y. and Kawada, T., 1995, Automotive air pollution in Jakarta with special emphasis on lead, particulate, and nitrogen dioxide, *Japanese Journal of Health and Human Ecology*, Tokyo, 61(5): 261–275.

12 Health and psychology of water

David Russell

Abstract

This chapter argues that decisions about water are too important to be left solely in the hands of the experts. Building on the recent research experience of the author, it is clear that the general public does not automatically trust government and/or industry to look after its health and environmental concerns over matters relating to water. The basis of this lack of trust is not only a degree of scepticism due to past experience but also has to do with the inherent duality of water; water being capable of bringing a vital resource to a community as well as of taking away its waste products. Underlying this surface concern about the state of our water are the more psychological or imagined aspects of water: its ability to evoke dreams and reveries that are vital for sustainability of culture. Involving the public in decision making is critical, not only because it will lead to better decisions, community building and ease of implementation, but just as importantly because, from a psychological perspective, it will foster the will to act.

Introduction

The character of social ecology has often been spoken of as being more adjectival than substantive. While as an emerging intellectual discipline it draws insights and strategies from the social and biological sciences, it is also an attitude and a desire. As an attitude, a social ecologist might seek to hold together, in dynamic tension, the ways to knowing, and doing, of science and the ways of knowing of the imagination. As a desire, these two cognitive orientations, the empirical world of sense perceptions and the imaginative world of image perceptions, come together as an intervention, an action project which represents a conscious design. A social ecology then can be understood as a perspective, a process, that can be designed into a building, a landscape, a work group, or (as is the focus of this chapter), into the production and use of reclaimed water.

Water as a scientific and cultural construct

Water, no less than the more general notions of 'landscape' and 'nature', is a social construction. Over time, society and, importantly, different subgroups of society (and for our purposes let us stay with scientists and poets) have invested particular meanings in 'water'. For the ecologist, water might be seen predominantly as a resource for plants and animals.

The focus for the plant scientist would be even more specific, namely, a raw material of photosynthesis. The poet, on the other hand, finds that water is the source of dreams and reveries. The cultural influence on how water is perceived is clearly visible in the case of the poetic imagination but is much more subtle when it comes to how water is appreciated scientifically. The philosopher Robin George Collingwood, who was a part-time archaeologist, gave us a useful image of the social construction of experience when he said that what turns up on your shovel depends on the quality of your imagination, of your knowledge, of your expectations.[1] What all this means is that we each have an investment, a stake, in how water is understood, how it is used, and how it is abused.

The historical duality of water

Joseph Campbell writes that

> water is the vehicle of the power of the goddess; but equally, it is she who personifies the mystery of the waters of birth and dissolution – whether of the individual or of the universe. . . the mystery of the origin of the 'great universe' or macrocosm; and the amniotic fluid is then precisely comparable to the water that in many mythologies, as well as in the pre-Socratic philosophy of the Greek sage Thales of Miletus (*c*. 640–546 BC), represents the elementary substance of all things.
>
> (Campbell, 1969: 64)

Most creation myths have water existing prior to the creation of earth. It was, as the first chapter of Genesis so eloquently expresses, the separation of the waters into those above and those below, that made space for any further creation. From this primordial separation comes the ubiquity of its dual nature. The duality of water is seen in its ability to purify, namely, to remove guilt and provide absolution, as well as to clean by removing dirt. Even its carrying capacity exhibits a duality; it both brings life-sustaining qualities and takes away our waste products. Water is found above the ground and below the ground. Water has a surface and pragmatic meaning as well as being a rich cultural source of imagery, metaphor and story. Gaston Bachelard, in his essay on the imagination of matter, *Water and Dreams*, speaks of water as an element that can 'engage the whole soul' and states that it can do this because it has the capacity for a 'dual participation of desire and fear, a participation of good and evil, a peaceful participation of black and white' (Bachelard, 1983: 12).

Historically, water has been asked to perform consistently this dual and seemingly contradictory function. It is the same river that is needed for our drinking water and for the disposal of human, industrial and agricultural waste. What one community puts into the river, another community further downstream will (more or less) have to contend with. While all water was judged to be in relationship to other water, the nature of the relationship was poorly understood until relatively recently. Ivan Illich traces the history of 'circulation' beginning with the ebbing and clumping and receding of blood in the body, the accepted view in medical practice until well into the eighteenth century. It was not until the mid-nineteenth century that the ideas of circulation were applied to the flow of water in an urban environment. Illich recalls the work of Edwin Chadwick who 'imagined the new city as a social body through which water must incessantly circulate, leaving it again as dirty sewage. . . . Unless water constantly circulates through the city, pumped in and channelled out, the interior space imagined by Chadwick can only stagnate and rot' (Illich, 1986: 45). As water was increasingly circulated in pipes and buried under the ground it was not only lost to view but also fell from consciousness. With the passage of time and the urbanisation of our daily living, all water is at risk of being reduced to functional H_2O. Water has gone into the closet!

Waste water and reclaimed water

As water has increasingly become a technology there has been an accompanying diminution of its inherent duality. As water has acquired a dominant utilitarian meaning it has progressively lost its psychological significance. This has not happened overnight. Water has been subject to a gradual transformation, especially over the past three hundred years. What we have now is water, largely conceptualised as a vehicle to do things with.

Most water is now consumed by industry and agriculture. Water has wonderful properties, of being a cleansing fluid, a collector of dirt or other contamination and of being able to carry the contamination away. But such a one-eyed view of the value and merits of water has been acquired at very high cost to the social and cultural well-being of a community.

This transformation is clearly apparent in the Hawkesbury–Nepean valley, the major catchment area for the city of Sydney. What were once mountain streams, flood plain creeks, and a meandering ever-changing river, are now a network of canals. These carry *predetermined* amounts of industrial and urban effluent as a managed commodity, into a river which has more the characteristics of a well-used bath tub, after being asked to manage many a wash in the same water.

Before dams and sewage effluent, this river was a coastal river prone to great flooding in times of heavy rain. In dry times, it would be a river bed that alternated between a creek and a robust fast-flowing river, especially as it crossed the flood plain. What we have today is a flow of sewage effluent and urban run-off, predictable other than in times of flood. The water level is constant but slight. The water laps up against the banks so that the image is more akin to a canal than to a river.

The white settlement of Terra Australis was characterised by a view of the earth in which 'science and capital were working together to apply scientific principles so that nature could be made to emulate industry' (Rosen, 1995: 27). It is not surprising then to find how quickly our major creeks and rivers were polluted in the name of 'improvement'. As early as 1796, the Sydney Tank Stream 'was so polluted as to be considered to be a danger to the health of the inhabitants (several of whom had died after imbibing it contents)' (ibid.: 19). As human occupation moved west into the Hawkesbury–Nepean valley the story was repeated. Wool-scouring plants and tanneries, coupled with the general disposal of slops, led to contamination of the Hawkesbury and South Creek. The editorial of *The Hawkesbury Chronicle and Farmers' Advocate* of 22 January 1887 not only complained about the tanneries but sought action regarding all the dead dogs and hogs in South Creek (ibid.: 93). Water was regarded as nothing more than a commodity that one could have concerns about.

What this means, psychologically, is that water has 'lost both its power to communicate by touch its deep-seated purity and its mystical power to wash off spiritual blemish. It has become an industrial and technical detergent, feared both as a poisonous stuff and as a corrosive for the skin' (Illich, 1986: 75).

The first appearance of a changing environmental consciousness emerged in the 1960s and continued to develop in the 1970s and 1980s. As a result of public demands for improved environmental management, sewage treatment facilities have improved but with only marginal impact on the quality of water in the river system. With the energy of government and society directed to the pursuit of a policy of minimally controlled urban development and the belief that they have the appropriate know-how to improve their technical 'management' of the waterways, the community is faced with an ever-increasing feeling of powerlessness. How can anything ever be done to fix up the river?

Public outcry has been followed by public cynicism as the relevant authorities have not been able to do anything other than provide a fig leaf for what the locals judge to be the obscenity of worsening pollution of this river and associated creeks.

It is the argument of this chapter that members of the community, because they are not dominated by a political or technological fixed mind-set, have a critical part to play in reclaiming our waterways and that their contribution will inevitably involve the ways of knowing of the imagination. Evidence for this 'inevitability' comes from a recognition of the cultural significance of water.[2] A rich symbolism which is beyond the surface of human consciousness is at the heart of the environmental movement's passionate concerns and the community's increasing anxiety about the state of its creeks and rivers.

It is timely to bring the two ways of knowing into conversation with each other. It is now obvious that after two hundred years of white settlement, the short-term and pragmatic mind-set alone does not contain the wisdom necessary for the task at hand. Out of this mind-set came the attitude of the early settlers that 'The river was an essential means of communication, a provider of food and water, yet the occupiers, rather than protect the resource, treated it as a garbage dump and sewer' (Rosen, 1995: 22). The descendants of these early settlers, by being open to the life of the imagination, may be able to see beyond pragmatic necessities. Thus they may recognise anew the psychological imperative of preserving rivers as places of soul, places for nourishment of the inner life.

Without such sources of soulful nourishment the risk of total cultural impoverishment is very real and such a loss would be too much for human society to bear.

Reclaiming our dreaming

In Hermann Hesse's novel *Siddhartha*, a young man in India spends a lifetime seeking knowledge – of life, of love, of longing and belonging. This knowledge is sought in vain, among religions, through self-denial, transcendence and intense study. Eventually, the knowledge is realised through worldly experience and finally by listening to the messages of a great river. Siddhartha's journey provided him with useful empirical knowledge but it was the river that conveyed wisdom. For water again to speak to us in our moment of need it must be brought 'out of the closet', out from the underground pipes, out of the tanks, and allowed to flow. Only in this way can we listen to the voice of water and look for what it is reflecting back to us.

All forms of water evoke different imaginings, images that extend far beyond utilitarian reality. Creek water, river water, meandering water, pond water, billabong water, cascading water, flood water – each has it own rich symbolism, each is embedded in a cultural history and, traditionally, each shows us new and useful aspects of death and rebirth.

The inherent duality of water makes it a powerful psychological symbol. For the imagination to become engaged some duality is essential. So we have duality as a characteristic of water both at the surface level of everyday events and at the deeper level of the imagination and reverie. As with all psychological duality, however, the opposites will involve each other, need each other, just as day needs night. While science seeks accurately to describe reality, it is the imagination that is 'the faculty for forming images of reality; it is the faculty for forming images that go beyond reality, which sing reality' (Bachelard, 1983: 16). It is also the imagination that allows for the 'transition from sensory values to sensual values' (ibid.: 20). It is precisely this transition which contains the motivation, the energy for action, especially action of a soulful kind.

Designing a social ecology of water

In all respects, reclaiming waste water by means of 'artificial' or constructed wetlands is a perfect opportunity for a socially ecological desire and attitude to find expression. Designing and managing such wetlands would allow for the dual participation that is so characteristic of water.

The participation of science ensures that the wetlands reach their potential effectively to transform urban run-off and sewage effluent into reclaimed water,[3] which becomes available again to the community for any use.

The participation of the public should be designed into the construction process to help ensure that *legitimate* decisions are made at the empirical level – the level of the senses. Not only does the decision itself need to be seen as legitimate – the process of decision making must fair, open to negotiation and based on best available data. There must be a sharing of *influence*, such that it can be seen that the public has had the opportunity to influence how the question or problem was posed, which alternatives were evaluated, and which adjustments were made to reduce impacts.

Dual participation needs to characterise the very process of public involvement if that participation is to possess a momentum capable of ensuring that proposed changes are carried through to completion.

There is a sensual consequence following the transformation of the sensible by the imagination. Given the cultural dominance of the *scientific* perspective, there is always a pressure or tendency for the *imaginal* to be down-played, omitted or ignored. A socially ecological perspective would reflect the importance of a process of designing the construction of wetlands, that enhances the function of those attributes and qualities that feed the human imagination.

Such a perspective encourages people to participate in a 'water mind-set' (Bachelard, 1983: 5–6). This approach is as essential as the equally important participatory involvement in critical decision-making. Both matters affect our daily living.

At one level, as residents of a community living with an urban wetland, we are concerned about health and lifestyle issues. Perhaps at another level we, the very same residents, are equally concerned about reclaiming water in a way that opens our imagination. This imagination has a quality of soul.

A constructed wetland, to be more than a 'pretence to the aesthetic symbol of a wedding between water and urban space' (Illich, 1986: 6), needs a design concept that is open to the imagination. Individual elements as much as the overall organisation needs to 'have soul'. Wetlands could be shaped as places where people wish to come and meditate on water. There could be ponds for reflection, reeds for listening to, stones and other forms for shaping and re-shaping the flow.

The story is told in Greek mythology that on a certain occasion, Pan had the temerity to compare his music with that of Apollo. At the subsequent challenge, only Midas judged Pan's music to be superior to the musicianship of Apollo. Apollo, the god of the lyre, was outraged and as a punishment for Midas, promptly transformed his depraved pair of ears into those of a donkey. King Midas tried to hide his misfortune under an ample turban, but his hairdresser found it too much for his discretion to keep such a secret; he dug a hole in the moist ground of a swamp and, stooping down, whispered the story into the ground, and covered it up. A thick bed of reeds springing up in the swamp began to whisper the story, and has continued to do so from that day to this, every time a breeze passes over the place (see Bulfinch, 1979, pp. 47–48).

What appears on the surface as an impossible error turns out, psychologically, to be a saving grace. The fact that a bed of reeds keeps the message alive for us today is at the heart of this story. James Hillman (1987) makes the distinction between the world of the spirit, symbolised by mountains, church spires and heroic quests, and the world of soul, the moist, fertile field of the valley and the wounded character. The donkey-eared world, the wounded world, is the imaginal world. The message of the reeds has a soul quality if we care to stop and listen.

Water has long been judged as being one of the 'hormones of the imagination' (Bachelard, 1983: vii). The very meditation on water 'cultivates an open imagination' (ibid.: 2).

A socially ecological design seeks to rehabilitate the imagination while at the same time rehabilitating sewage effluent and reclaiming urban run-off.

Conclusion

Our culture has pushed the imaginative into being peripheral, trivial to the pursuit of empirical knowledge, commerce and engineering. Yet it is the imagination that embodies the energy to 'make a difference'. We have a dualistic separation that divides us from ourselves, from our sources of energy, from creative processes. The firm ground of empirical science has won over the ever shifting, quite mercurial, base of the imagination. In doing so we have largely lost our ability to explore the conversation of imagination and are limited to the conversation of the senses. Yet it is the imagination that carries us into conversational familiarity with the physical and biological environments that we inhabit. How much better to take heart in both domains.

The challenge is to design a framework that will constitute an over-arching ecology, one that interconnects the pattern of events and the pattern of images. Such is the challenge of designing a social ecology!

Notes

1 This reference to Collingwood was found in Skolimowski, *The Participatory Mind*, p. 394.
2 The author had a major involvement (1992–1994) in a research project based in the Hawkesbury valley which had as one of its aims the development of strategies for the greater involvement of the local residents in decisions to do with water and waste-water management. During the course of this work it became apparent that residents were sceptical of government agency commitment to bring about changes that would significantly improve the state of the catchment. At the same time, it was apparent that this cross-section of residents has a great desire to talk at length about their feelings for the river including memories of past experiences and the hope that their children, or at least their grandchildren, would not be deprived of such experiences.
3 Artificial wetlands 'have the potential to achieve high removal of BOD (biochemical oxygen demand: a measure of the availability of material as a biological food and by the micro-organisms during oxidation) suspended solids, nutrients, trace organics, and heavy metals. Pathogenic bacteria are removed via sedimentation, predation, and natural die-off. Pathogenic die-off due to natural UV radiation occurs in artificial wetlands containing open water sections. Significant die-off rates have also been demonstrated for viruses.' *Draft Guidelines for Sewerage Systems – Use of Reclaimed Water*, in the National Water Quality Management Strategy, No. 14, Commonwealth of Australia, April 1996, pp. 13–14.

References

Bachelard, G., 1983, *Water and Dreams: An Essay on the Imagination of Matter*, Pegasus Foundation, Dallas.
Bulfinch, T., 1979, *Bulfinch's Mythology*, Avenel Books, Avenel, NJ.

Campbell, J., 1969, *The Masks of God: Primitive Mythology*, Penguin, Middlessex.

Hesse, H., 1973, *Siddhartha*, Picador, London.

Hillman, J., 1987, *Peaks and Vales*, Puer Papers, Spring, Dallas.

Illich, I., 1986, H_2O *and the Waters of Forgetfulness*, Marion Boyars, London.

National Water Quality Management Strategy, 1996, *Draft Guidelines for Sewerage Systems – Use of Reclaimed Water*, No. 14, Commonwealth of Australia, April.

Rosen, S., 1995, *Losing Ground: An Environmental History of the Hawkesbury–Nepean Catchment*, Hale & Iremonger, Sydney.

Skolimowski, H., 1994, *The Participatory Mind*, Penguin, London.

13 Health Impact Assessment in Flanders

Contribution to environmental management and health

Peter Janssens and Luc Hens

Abstract

Environmental Impact Assessments (EIAs) performed so far in Flanders, Belgium, differ widely in their content and approach. To lessen the problems, the Flemish government has decided to develop a manual of guidelines for EIA.

A methodology has been proposed for each of nine knowledge areas or disciplines: Air, Climate, Noise and vibration, Radiation, Soil, Water, Fauna and flora, Landscape and monuments, and People (health and population density). This chapter summarises the methodology dealing with the health aspects of EIA.

As health is intrinsically related to the environment, Environmental Health Impact Assessment (EHIA) fundamentally rests on the description of environmental changes. Health risk assessment uses toxicological and epidemiological methods. Using these results, the expert predicts the health effects of environmental changes associated with a project or activity.

Introduction

In Belgium, Environmental Impact Assessment (EIA) is regionalised. Belgium is divided into three regions: Brussels, Wallonia and Flanders. Legislation for EIA in Flanders differs from that in the other regions. Devuyst and his colleagues have evaluated this situation (Devuyst *et al.*, 1991a, 1991b, 1993). Since 1989 the Flemish regional government has identified a list of projects and activities which should be subject to EIA (Table 13.1).

Following the introduction of EIA legislation in the Flanders region, a uniform format and framework of methodology and presentation between various EIA reports has yet to be developed. The contents and methodology used in the assessments developed to date differ markedly.

A description and evaluation of EIA in the Flanders region was performed by Devuyst and colleagues (1993). This study revealed a lack of quality control concerning the content of Environmental Impact Statements (EISs).

To facilitate a resolution of this difficulty, the Flemish regional government determined to develop benchmarks and publish guidelines for EIA. The primary goal of these is to establish a basis on which experts can rely when formulating EISs.

Table 13.1 Overview of projects subject to Environmental Impact Assessment in Flanders, Belgium

Infrastructure projects	Sources of contamination
Highways and railways	Animals
(construction and major changes)	Poultry-rearing installations
Airports	Pig-rearing installations
(construction and major changes)	Installations for rearing fur-bearing
Ports	animals
(construction and major changes)	Installations for rearing native
Seaports	mammals
Yacht marinas	Chemical industry
Ports for inland-waterway traffic	Integrated chemical installations
Inland waterways	Production of phenol, pesticides,
Industrial activities	amines
Urban development	Crude oil refineries
Holiday villages and hotel	Waste
complexes	Incinerators and processing
Construction of permanent	Storage of dangerous and toxic waste
recreation areas	Landfills
Golf courses	Major industrial installations
Pipeline installations	Gasification of coal
Pipeline installations in sensitive	Smelting-furnaces and steelworks
areas	Processing of ores
Major pipeline installations for	Production of non-ferrous metals
drinking water	Energy industry
Water winning	Thermal power stations
Large water basins	Transformation to other fuels of
Water management	thermal power stations
Land consolidation and planning	Installations for processing and
Deforestation and initial reforestation	transformation of asbestos
	Major extraction sites and mining
	activities
	Storage and trans-shipment of liquid
	natural gas (LNG)
	Storage and trans-shipment of coal and
	minerals

Sources: Devuyst *et al.*, 1991; Ministerie van de Vlaamse Gemeenschap, 1989.

These guidelines will also provide the Administration for Environment, Nature and Land Use (AMINAL) with an instrument for evaluating EISs. In addition, appropriately detailed summaries of these guidelines aim to provide information to relevant target groups of the EIA process such as initiators, concerned environmental Non-Governmental Organisations (NGOs), and interested individuals.

This chapter has three parts. The first describes how EIA should be performed in Flanders and provides an overview of relevant legislations. The second considers terminology and the methodologies to be used in the assessment of environmental impacts from the viewpoint of the different EIA disciplines. These categories are: Air, Climate, Noise and vibration, Radiation, Soil, Water, Fauna and flora, Landscape and monuments, and People (health and population density).

Based on a European Community (EC) directive of 1985, the administration appoints experts for each of these EIA disciplines. The work of these experts is supervised by a coordinator. The third section independently considers project groupings (Table 13.1). It explains the specific needs and difficulties of assessing environmental impacts of these project categories.

The chapter reviews the general methodology for those guidelines concerning health impact assessment, as well as general concepts regarding health dimensions of EHIA. When we investigate the importance of EHIA in other countries, we find that research in this field has only very recently become a part of EIA procedures.

Countries that took a leading role in the promotion of EIA currently have the most complex system for EHIA. In the United States the National Technical Information Service (NTIS) of the Department of Commerce has published a theoretical guide for assessing the health effects of environmental changes, the *Public Health Assessment Guidance Manual*. The Environment Protection Agency (EPA) has also published a large number of practical guides. These concentrate on specific industries or other contaminating activities.

In the Netherlands much research has been devoted to the composition of published guides for different disciplines, including health. Additional references are available offering information concerning specific contaminating activities.

Guidelines for Environmental Health Impact Assessment in Flanders

Because people are in constant interaction with their environment, their health and wellbeing is very closely connected with the quality of that environment. Environmental transformations may result in disease and discomfort, and EHIA should therefore be an integral part of EIA. Because of these interrelations of environment and health, the identification, description and evaluation of potential harmful factors is fundamental to EHIA. Table 13.2 reviews the six basic steps of EHIA.

The assessment of health risks needs a holistic framework in order to construct an integrated report. Before integrative disciplines can be employed, technical disciplines have to be applied to yield necessary information. This requires a constructive and interactive dialogue as the central form of communication between these technical and integrative disciplines. Such fundamental concepts must be built into all phases of EIA, both before and during project planning.

Elements of the process of EHIA

Description of the project site

Detailed description of the project site is an essential element of EHIA. A visit to the site is indispensable in order to establish an overview of the project setting and its surroundings. These surroundings have many dimensions that are crucial to the study of possible health effects consequent to changes in the environment.

Table 13.2 Six steps of Environmental Health Impact Assessment

1	Description of the project site
2	Identification of possible changes in the environment relevant to the health of the population
3	Identification and quantification of the exposure and dose
4	Prediction of the relevant health effects
5	Evaluation of those predictions
6	Identification and evaluation of mitigating measures

The dimensions of an area studied depend on many factors. These include:

- features of that project such as material and processes utilised as well as the size and location of the project;
- features of the surroundings: topography, geology, climatic conditions, soil characteristics, history of the site, etc.;
- features of the human population: demography, socio-economic status, land use, utilisation of natural resources, health status, pattern of activities, individual differences, etc. (ATSDR, 1992).

After consideration of these elements it should be possible to identify at-risk groups, which deserve special attention (although, of course, this does not mean that other sub-populations should be ignored).

The content and scope of the EIA will depend largely on this detailed project description. This will determine the size of the studied area and which specific factors should receive closer attention.

For the EHIA in Flanders we decided to follow a subject-discipline approach. Classifying the area under investigation according to its geography is an important step in EIA. In this way the area for study can be physically delineated so that information gathered and consequent research can concentrate on specific aspects. In order to accomplish this, experts need an extensive description of the project and its surroundings including landscape and residents.

Identification of possible changes in the environment relevant to the health of the population

The project description must provide experts with sufficient information to establish an overview of possible environmental changes which may occur when the project is carried out. Using this description, it is possible to identify and minimise those changes capable of affecting the health of residents.

Possible environmental changes should not be limited to chemical substance release into the environment. Physical factors such as noise and vibration as well as biological factors including micro-organisms and vermin should also be considered.

Threats to health need to be interpreted very broadly. Not only should disease and discomfort be considered, but possible changes in psychological and social well-being should also be part of the discussion. It is, however, very difficult to quantify discomfort, nuisance or psychological effects.

Identification and quantification of the exposure pathways and dose

The pathways of factors capable of causing adverse health effects to people can be followed and mapped through the environment, and the process starts with the production and emission of the compound in question. In their passage through air, water and soil, the initial compound may undergo many changes such as dilution, transformation and accumulation.

The impact of exposure depends not only on the external concentration in the environment, but also on individual personal factors such as age, gender, diet, and psychological and physical health.

When evaluating exposure, one has to take into account not only pollution caused by the project, but also already established pollution from other activities, either past or present. The history of the project site makes possible the identification of eventual, actual or prior pollution.

In all organisms, chemical compounds with toxic potential follow similar pathways: uptake (inhalation, ingestion, physical contact), transport, bio-transformation, eventual accumulation and excretion. Somewhere along these pathways, compounds will eventually reach their 'target' and exert their effect. Ideally it should be possible to measure the dose at the level of the target, cell, tissue or organ. In most cases this is not feasible, so models and estimations have to be used.

Prediction of relevant specific health impacts of a project

Prediction of the probability and seriousness of a health effect is the main aim of EHIA. This is underpinned by toxicological and epidemiological research. The goal of such research is to establish a relation between dose and effect.

Such data very often result from animal experiments. Interpretation of the results necessitates extrapolation. This process is characterised by considerable uncertainty and calls for risk analysis to estimate a margin of safety and ongoing monitoring. Data on human subjects occasionally becomes available as a result of chronic occupational or acute accidental exposure to high doses of a substance. EHIA needs data from chronic, low dose exposures to make reliable predictions concerning the health impact of a project. More extrapolation is necessary when these data are not available, increasing the uncertainty of the predictions.

A major difficulty faced by environmental toxicology is the health-risk assessment of mixtures of chemicals, so-called cocktails. The health risk of such cocktails can be lower (antagonism) or higher (synergism) than the summation of the individual risks. Synergism in particular poses serious problems in health risk assessment. A well-known example of synergism is the increase of cancer risk due to exposure to asbestos when the exposed individuals are regular tobacco smokers.

Table 13.3 Zaponi list of data for use in evaluating the health impact of
chemical substances

1. Name and code of the chemical substance (EG-code, or any other code). This
   can be completed with the chemical formula and eventually with current
   synonyms.
2. Acute toxicity (LD_{50}) classification. Possible effects.
3. Sub-acute toxicity, NOEL. Possible effects.
4. Chronic toxicity, NOEL. Possible effects.
5. Carcinogenicity (yes/no, IARC or EPA classification)
6. Mutagenicity (yes/no, used test)
7. Teratogenicity (yes/no, used test, NOEL)
8. Corrosiveness, irritation (classification)
9. Inflammability
10. Danger of explosion
11. Available data concerning occupational risk (lowest dose that causes adverse health
    effects, doses without observed health effect, classification of the effects)
12. Available data concerning epidemiological studies and risks for the total population
    and specific sub-populations (see above)
13. Guidelines, standards (general population, occupational situations, environmental
    quality standards, other available standards)
14. Persistence in the environment (half-life in different environmental compartments,
    rate of breakdown, other persistence indices). Orders of magnitude are generally
    sufficient.
15. Water/air-distribution coefficient
16. Soil/water-distribution coefficient
17. Degree of bio-accumulation (directly measured or estimated). If directly measured
    the biological species should be mentioned. If estimated the method should be
    mentioned.
18. Any other relevant information concerning substances

Source: Zaponi, 1989

Epidemiological studies try to relate observations in existing populations to
their environmental conditions. Care has to be taken not to be misled by spurious
correlation. A numerical relation between two events does not necessarily mean
that there is a causal link between them. Because of the brief time frame for most
Environmental Impact Assessments, cross-sectional epidemiological studies are
usually recommended.

The results of the toxicological and epidemiological studies can be summarised
in a conveniently arranged table. Zaponi (1989) suggested a list of data useful for
an EHIA of chemical substances. Several substances may be examined during the
study of most projects. Relevant data from the Zaponi list can be combined in a
matrix. In this way an overview can be offered of the available information relevant for health risk analysis.

Evaluation of the health impact predictions for a project

The results from the investigations described above have to be interpreted. For
many reasons, the interpretation of seemingly similar data can lead to very different

outcomes, but the procedure is nevertheless essential as scientific results usually display statistical dispersion. The opinion of expert evaluators should be available for inclusion.

The health effects of a project can be characterised in two ways:

- nature of the effect: acute or chronic, reversible or irreversible, local or systemic, stochastic or non-stochastic;
- nature of the target: nervous system, liver, kidneys, respiratory tract, reproductive system.

The occurrence of cancer should receive special attention. It is a typical stochastic effect. The extent of the effect is not dose dependent; the probability of effect, on the other hand, is. In most cases there is no threshold. Non-stochastic effects usually have a threshold. The intensity of the effect is dose dependent.

EHIA is developed by comparing two situations:

- The reference situation is usually the actual present setting. This description will concentrate only on relevant environmental factors.
- The planned situation is the predictive description of the environment once the project is actualised and operational.

During the EHIA, to evaluate the health impact of a project both the actual (reference) and planned situations should be described in as much detail as possible. With the actual (reference) situation in mind, the expert tries to predict and evaluate the health effects. The predicted effects are grouped to facilitate the coherent discussion and evaluation of the project. The process of evaluation should sequentially reveal how extensive health impact effects will be in the planned situation.

Every project has potential positive and negative impacts, entailing the health status of people. The task of the EHIA is to weigh up the pros and cons and choose that option with the most minimal health burden.

Identification and evaluation of mitigating measures of a project

All activities have impacts on their environment. In the process of EIA, alternatives are evaluated and the most appropriate one selected. Such selection is a compromise between possible impacts at different levels. The selected alternative may carry some health burden.

EHIA, therefore, should be an integral part of the EIA process. This means that many other factors will be considered. Following the integration of all conceivable aspects and evaluation of all possible effects, EIA experts have to propose their best alternative, which may not necessarily be the best from the point of view of the health assessor.

The primary goal of EHIA is to avoid or minimise potential adverse health effects of projects. Projects are examined before actualisation. This preventive approach allows the government to take appropriate actions in order to protect the environment and the health of the people. Once health risks are assessed, it is possible to recommend mitigating measures to minimise those risks. The recommended measures should be evaluated for their feasibility. Best available techniques not entailing excessive costs (BATNEEC) should be applied.

Post-monitoring and post-evaluation stages

After monitoring and evaluation, the project is followed during planning, preparation, construction and exploitation (and eventually during demolition). The observed effects are then compared with those predicted in the Environmental Impact Statements. This evaluation offers an opportunity to refine the methodology of prediction and allows experts to obtain new data about substances whose impacts are not well known.

Conclusion

EHIA studies the health impact of environmental changes in EIA. This central aspect continues to be frequently neglected. Published guidelines should provide EIA professionals with a suitable framework to consider health risk assessment. In this way the Flemish regional government wishes to encourage the introduction of continuity and uniformity in the collection and presentation of data and the consequent of conclusions.

EHIA faces many practical problems. Its primary data rely on other, more technical, disciplines. Health impact prediction has to rely on scientific research. Data concerning many substances are often scarce or non-existent. Data that do exist will show variations typical of statistical observation. EHIA experts must also take into account extensive individual variations resulting from genetic variability and personal behaviour. The consequent results, statistically unreliable though they are, must not be neglected or ignored.

Important difficulties also stem from the interdisciplinary character of EHIA. On the one hand, EIA experts are most often discipline-oriented and consequently the transdisciplinary aspects of an EIS are often its weakest part. On the other hand, this lack of an interdisciplinary approach makes the field particularly attractive to human ecology researchers.

The creation and review of these frameworks and guidelines should be a dynamic process. Once completed, the material should be evaluated and tested in a working situation. In this way the guidelines can be confirmed from the point of view of their value in practice. During this process the proposed methodology can be confirmed and further adapted. There is a need for international and transdisciplinary collaboration in this area.

Using this framework, AMINAL proposes to test these guidelines in a specific case study. They will be evaluated in an existing situation. The outcome should be a realistic and practical workbook.

References

Agency for Toxic Substances and Disease Registry (ATSDR), 1992, *Public Health Assessment Guidance Manual*, US Department of Health and Human Services, Atlanta.

Arquiaga, M.C., Canter, L.W. and Nelson, D.I., 1994, Integration of health impact considerations in environmental impact studies, *Impact Assessment*, Vol. 12, No. 2, pp. 175–97.

Devuyst, D. and Hens, L., 1991a, Environmental impact assessment in Belgium, *Environmental Impact Assessment Review*, Vol. 11, No. 2, pp. 157–69.

Devuyst, D. and Hens, L., 1991b, Environmental impact assessment in Belgium: An overview, *The Environmental Professional*, Vol. 13, pp. 166–73.

Devuyst, D., Nierynck, E., Hens, L., Ceuterick, D., De Baere, V. and Wouters, G., 1993, Environmental impact assessment in Flanders, Belgium: An evaluation of the administrative procedure, *Environmental Management*, Vol. 17, No. 3, pp. 395–408.

Ministerie van de Vlaamse Gemeenschap, 1989, *Milieu-effect rapportering in het Vlaamse Gewest*, Report number D/1989/3241/25, Brussels.

Turnbull, R.G.H. (ed.), 1992, *Environmental and health impact assessment of development projects*, A handbook for practitioners, Elsevier, London.

Zaponi G.A., 1989, Methods for the health component of EIA of industrial development projects, *Proceedings of the 10th International Seminar on EIA and Management*, 9–22 July, University of Aberdeen, Scotland.

Index

abortions 165
accidents involving children 196
acclimation, definition 107
acclimatisation, definition 107
acclimatory adjustment 87; definition 107
accountability 49
Acock, A. and Kiecolt, K.J. on divorce 198
adaptability 48, 85–7; definition 107
adaptation capabilities 87–91, 106; definition 107; of organisms 105–6
adaptational changes: of cardio-respiratory functions 89; and stress 88
adaptational diseases 106
adaptive behaviour, and modernisation 90
Adelaide Recommendations 37, 49
adjustment, definition 107
adolescent offending 185–7
advertising, and concept of beauty 170
Africa, traditional health systems 213–14
age: effect of 80; and sexual activity 159
ageing population 44
AIDS, and traditional medicine 213–14
air pollution, Jakarta 243–8
alcoholic patients, and traditional medicine 213
alienation 194
alternative medicine 209–10, see also traditional health systems; traditional medicines
America see USA
Americas, traditional health systems 212–13
anthropo-pressure 105
anthropocentrism 46
anthropology, ecological 46
Antonovski, A. 46
applied science, and basic science 66
artificial fertilisation 159–60
Asia, traditional medicine 211–12

aspirations 87; and human beings 81; and population growth 69; and technology 72
assessments, shortcomings of 122
Australia 51; environmental research 66; National Health and Medical Research Council 74–5; population 69, 72–3; review of strategic planning 121; sustainable society 62, 63; water 252–3
Australian National Strategy for Ecologically Sustainable Development 117

Bachelard, Gaston on water 251
balance 83
Bangladesh, traditional medicine 212
Barrow, C.J. on carrying capacity 125
Barrow, H.H. on human ecology 7
Bartley, M. see Smith, G.D. et al.
Bateson, G.: adaptability 48; ecology 56–7; patterns 45; tyranny and flexibility 56; unconnected speculation 47
Baudot, P. 169
Beakhurst, Graham, political ecology 53
beauty, and advertising 170
behaviour, adaptive 90
Belgium see Flanders
beliefs, USA 137
bio-logic 8
biodiversity: and health 209–10; need to prevent further loss of 118–19; policies 60–61, 224; and traditional medicines 218–23
biodiversity prospecting 215
Biodiversity Support Programme 221–2
biological engineering 5
biological reference systems 94, 97–8

biological status (human), monitoring of
96, 98–101
biological tolerance, and climatic zones 84
biosphere, human use of 60
biotic balance 4–5
biotic systems, and change 63–4
birth control 155; and increase in
menstrual cycles 157; methods 166–9,
see also contraception
birth rates and death rates: population
increase 230–31
Bishaw, M. on traditional medicine 219
Blackman, T., Evason, E., Melaugh, M.
and Woods, R. on accidents involving
children 196
Blane, D. *see* Smith, G.D. *et al.*
Bley, D. and Baudot, P. on efficiency of
contraceptive pill 169
blood *see* haemoglobin level
breast cancer 157
Brewer, G.D. on ecological public health
46
Bronfenbrenner, U. 47
Brundtland Report 38, 42, 53, 55, *see also*
Our Common Future; World
Commission on Environment and
Development
Bubloz, M.M. on families 18
budgets, and planning 130
Butler, N.R. 177–8

Caldwell, L.K., sanative communities 74
Cameroon, over-harvesting of medicinal
plants 221
Campbell, Joseph on water 251
Canada, and traditional medicine 213
capacitance, definition 107–8
capital stock, necessary for sustainable
society 73–4
Carpenter, W. on holistic approach 12;
liberal arts 14
carrying capacity 125
Centre for Human Settlements xvii
Chamberlin, R.W. 47; public health
strategy 50–51
change: biotic systems 63–4;
compensation 83; environmental 92;
political pressure for 51
chaos theory 46
chauvinism, eschewed by human ecology 8
Chiang Mai Declaration 220, 221
child abuse, and homelessness 203
child health: causal relationships in 187–8;
and community resources 51;

developing and developed countries
compared 179; education 176, 177;
hygiene 176, 180; inter-generational
effects 181; macro and micro
environmental explanations 175–89;
poverty 178–9, 180, 181, 182; smoking
179–80
child mortality 183, *see also* infant
mortality
children: and accidents 196; development
of 94; and divorce 198
China: population control 232; and
traditional medicine 212, 220–21
Chivian, E. on danger to biodiversity
118–19
cigarettes, and the media 145
civilisation-affected diseases 104–7
civilisations, past and present 106
climatic zones, and biological tolerance
84
Club of Rome 43
colonialism 228–30
commercial advertising, and self-
perception 170
commercial interests, and media 144–5
Committee on Nutrition xvii, *see also*
Commonwealth Human Ecology
Council (CHEC)
Commonwealth Human Ecology Council
(CHEC), programmes xv, xvi, xvii-xviii
Communist Manifesto 40
communities 18–19; and health 21,
50–51; and human ecology 9
community participation in planning
201
compensation, and change 83
complementary medicine 209–10, *see also*
traditional health systems; traditional
medicines
compliance, and punishment 70
conflict in families 199
conservation, and health 59–75
consumer behaviour 52
contraception: efficiency of pill 169;
impact in Spain 164; traditional 169, *see*
also birth control
'Contract with America' 136, 138–9,
147
contractual relationships 196–8
Cortese, A.D. on health and
environmental crisis 118–19
crisis driven policies 61–2
cultural adaptation, definition 107
cultural adjustment, definition 107

cultural factors 19, 83, 154; infant
mortality 104; meaning of 67–8;
traditional medicine 211; values and
human population issues 69

Daysh, Zena: history of human ecology 4;
sustainability 12–13
death rates, and population increases
230–31
deaths, attributable to pollution 118
decision-making: new form required 53–4;
and planners 120
degradation, and adaptation 106
deism 12
democracy, importance of 71
demographic winter 69
Denmark, homelessness 203
deprivation, and child health 180
developed countries: fertility rates 230;
and public health concerns 37
developing countries: child health 178–9;
destruction of environment 131; and
development 228; infectious diseases
155; population increase 230–32;
sustainable development 118; traditional
health care 211–15
development: and free market mechanisms
113; health concerns 70; monitoring
impact 70–1; threshold to 127; welfare
improvements 112
developmental adjustment 87; definition
107
developmental homeostasis 79
dieting 170
disease xvi; causes of 92; civilisation-
affected 104–7; ecological framework
153
distal causes of child ill-health 176
distress 106–7; definition 108
distribution in biotic systems 63
diversity 64
divorce 198
Dolanski, E.A. *see* Frisch, A.S. *et al.*
domestic animals 69
domestic violence 198
Drewnowski, J. 117
drinking water: Indonesia 237–8, *see also*
water
Dwivedi, O.P. 53
dysfunctional parenting, and socio-
economic status 180

Earth Summit, Rio de Janeiro 6–7, 119
eco-logic 8

eco-sensitivity 85–6
ecological principle, and sewerage principle
43–4
ecological public health, theory of 45–7
ecological regularity 85
ecology 2–4, 46; and environmentalism
65–6; holistic approach 3; and humans
61; Java 232–5; and public health 37–8,
44, 45–7, 52–7, *see also* human ecology;
political ecology
economic crisis 6; and child health 181;
and infant mortality 101
economic growth: and health 40; limits to
113; and progress 38–9; and traditional
medicines 222–3
economics 7
ecosystems 3, 8, 14; concept of and
human ecology 10; definition 154; and
health 130; human 16–17
education 40; child health 176, 177;
fertility 165–6; and health 100; and
parenting 180; population increases
231–2; traditional medicines 224
electronic data bases, use in medical
research 217
empowerment, and information 73
Endod 219
energy crises 6
engineering, biological and social 5
entropy 80
environment 82–5; constraints 124–6;
definition 8; degradation of 51, 71; and
ecology 3, 65–6; and genetic
determinants 93–4; and health 1, 82–5,
114, 118–19, 261–4; and medicine
81–2; and organism: relationship
85–92; and people 67–9; problems of
and poverty 69; research into: flaws
66
Environmental Impact Assessments
(EIAs): Flanders 258–65; shortcomings
of 121–2
environmental resources, rational use of
124–5
environmental threshold 127–30
environmentalism, and ecology 65–6
epidemiologic transition theory 155
equilibrium, maintenance of 87
ethical framework 64–5
ethical systems, and poverty 104
Ethiopia, and research into traditional
medicines 219
ethnobotanical research 213
Europe, infant mortality 103

European Charter on Environment and Health 59, 74
European Home and Leisure Accident Surveillance System 195
eutrophication, and fertilisers 242
Evason, E. *see* Blackman *et al.*
evo-deviations 92; definition 108
evolutionary wisdom 105
expediency, and planning 121
exposure, evaluation of 262
extinction of species, high rate of 118–19

factors: exogenous 82–5, *see also* modifiers
faecal waste, and child health 179
families 18; and conflict 199, 204; and healthy homes 196–9; and human ecology 9; and the individual 101; sizes 161
family factors 83
family income (USA), changes in 139
family planning 40, 230, 231, 232
fecundity *see* fertility
female wage labour, cheap 199
feminisation of poverty 47
feminist analysis of societal values 38
fertile life, variability 163
fertilisers, and eutrophication 242
fertility: biological and social constraints on 156; changes in Spain 164–5; Java 235; and socio-economic variations 165–6, *see also* reproductive health
fertility control, Morocco 161
fertility rate: developed countries 230; and maternal mortality 169
Flanders 258–65
folk 5
food 93
food chain, and toxic chemicals 54
forest management, Indonesia 234
free market mechanisms, and development problems 113
Frisch, A.S., Kallen, D.J., Griffore, R.J. and Dolanski, E.A. on infant survival 184–5
funding traditional health care 214–15, 224

Gaussian distribution 100
general systems concepts, and human ecology 10
genetic adaptation 85; definition 107, *see also* phylogenesis
genetic determinants, and environment 93–4

genetic predispositions 81
genotype, definition 108
geographic environment, and development 127–30
geography and human ecology 7
Germany, herbal medicine 220
Giddens, A. on the home 198
Global Biodiversity Strategy 60
Global Strategy for Health for All by the Year 2000: 41–2, 48–9, *see also* World Health Organisation
Golding, J. and Butler, N.R., on causes of infant mortality 177–8
government, US public opinion of 138
Greek mythology, and water 255
Green, J.L., economics 7
Griffore, R.J. *see* Frisch, A.S. *et al.*
gross domestic product, limitations as measure of national wealth 70
gross national product: infant mortality in Poland 102; life expectancy 229, 230; and progress 38; public health 56
growth, and sustainable development 112

habitat 81, 82, *see also* environment
Hackenberg, R. 155
Haeckel, Ernst on ecology 2
haemoglobin level: hematocrit index 98; as index of health 97
Hancock, Trevor on need for sustainable policies 54
Hart, N. on marital breakdown 198, 199
Hawley, Amos on human ecology 5
health xvi, 19–24, 70, 119, 209; access to 47; overview 79–82; positive and negative indices 94–104; as social right 52, *see also* traditional health systems
health care reform: debate 142; USA 146
health ecology, definition 22–3
health field concept 45
health promotion, and campaign against smoking 179–80, 187–8
health risk patterns, societal response 39–41
hematocrit index, and haemoglobin level 98
herbal medicines: Germany 221, *see also* traditional health systems; traditional medicines
Hesse, Herman 254
heterogeneous populations 93
heterosis, definition 108
heterozygosity and homozygosity 93

heterozygous, definition 108
Hillman, James 255
holistic approach: and ecology 3, 8,
    11–13; health xvii, 1–2, 37, 210; shift of
    emphasis away from 41
homelessness 194, 202–4
homeorhesis 79, 105; definition 108
homeostasis 80, 105; definition 108
homes: accidents in 195–6; and the family
    196–9; and health 193–6; and
    homelessness 202–4; and the
    neighbourhood 199–201, *see also*
    housing
homogenousness, definition 108
homosis, definition 108
homozygosity and heterozygosity 93
homozygous, definition 108
housing 40; and health 195–6, *see also*
    homes
human biology, definition 108
human ecology 46; applications of 15–16;
    and chauvinism 8; and community 9;
    definition xv-xvi, 7–11, 108; history
    4–7; holistic approach to 3, 8, 11–13;
    and individuals 17; values 13–15, *see
    also* ecology
human ecosystems 16–17; structural
    components of 154
human interaction, ecology of 46
human needs, categorisation of 116
human settlements, aims of 117
human-logic 8
humanism, and human ecology 9
humans: aspirations 81; concepts
    concerning 10–11, *see also* human
    ecology
Huxley, Julian, biological and social
    engineering 5
hygiene, and child health 176, 180
hysterectomies, and menstrual disorders
    157

Illich, Ivan 252
impact of new development, monitoring of
    70–1
income, and living conditions 94–5
India: homelessness 202; supply of
    medicinal plants 220; traditional
    medicine 211–12
individuals: and the family 101; and health
    20–1; and human ecology 17
Indonesia 227–8; air pollution 243–8;
    drinking water 237–8; irrigation
    239–41; Java 232–5; population density

234; sewage treatment 241–2; washing
    water 239; water cycle 235–7; and water
    quality 242–3; well water 238–9
industrialisation, China 232
industrialised societies, and public health
    38, 97, 261–4
industry: need to improve regulatory
    mechanisms 70; public concern over
    waste 199–200
inequality, and domestic violence 198
infant mortality 184; and culture 101–4;
    Europe 103; Poland 102, 104; and
    social class 177–8; Sri Lanka 184, *see
    also* child mortality
infant survival in intensive care 184–5
infectious disease pattern, developing
    countries 155
information: and empowerment 73; and
    the media 142–4; and policy
    formulation 71; and public opinion
    139
information access 64
information sources, USA 140–1
inner growth xvi
input-output factors 17
institutions, ossification 65
integrated planning 122
integration of disciplines 7
intellectual property rights, and traditional
    medicines 218–19, 224
Inter-departmental Committee (UK), on
    education 177
inter-disciplinary nature of human ecology
    8
inter-generational effects, child health 181
interactional outcomes, human beings 11
interdependence 17
International Conference on Health
    Promotion 42, *see also* Ottawa Charter
    for Health Promotion
International Conference on Population
    and Development 155–6
International Human Ecology Conference,
    Tokyo xvi
International Monetary Fund 39; and
    gross national product 56
interrelations, science of 61
irrigation, Indonesia 239–41
isolation of neighbourhoods 200

Jakarta, air pollution 243–8
Japan: lead pollution 246–7; rice
    cultivation 242
Java, fertility 235

Jewish culture, and infant mortality in Poland 104
Johnson on problem solving 15–16
juvenile crime *see* adolescent offending

Kallen, D.J. *see* Frisch, A.S. *et al.*
Kickbusch, Ilona on public health 115
Kiecolt, K.J. 198
Korea, and traditional medicine 212
Kozlowski, J.: on threshold to development 127; on Ultimate Environmental Threshold 129
Kuhn, T., paradigm shift 47

Lalonde Report 153
land available per caput 69
language, and rules 195
Latin American Centre for Social Ecology (CLAES) 6
Leach, E.R. on conflict in marriage 198
lead pollution 246–7
Lévi-Strauss, Claude 228
liberal arts 14
life expectancy 23; and gross national product 229, 230
living conditions: improvements in 39; and income 94–5; need to improve 42

McKeown, T. 54; and gross national product 56; on public health and medical science 39–40
maladjustment, definition 108
malaria, and traditional medicine 214, 215
Malawi, and traditional medicine 214
Malta xvii
Man and the Biosphere Programme, UNESCO 73
management, and ecology 65
mandala of health 37
marriage: breakdown 198, 199; and conflict 198; and maternity 164–5; variation in age 165
maternal death rate 169, 170
maternity 159–60; first 171; and marriage 164–5
media: and commercial interests 145; and public information 143–4; USA 142–7
medical profession: and public health 47; and reproductive health 170
medical research, electronic data bases 217
medical science, and public health 39–40
medicinal plants, endangered 210, 220–23

medicine: and environment 81–2; expediency and biodiversity 118–19; and health 42, 193; limitations of 53; and safety 216–17; traditional and modern compared 211, *see also* traditional health systems; traditional medicines
Melaugh, M. *see* Blackman, T. *et al.*
menarche 157–9, 171, *see also* fertility; reproductive health
menopause 158–60
menstrual cycles, numbers during fertile life phase 157, 161, 164
menstrual dysfunctions, and early menarche 157
mental illness, and homelessness 203, 204
Mexico, and traditional medicine 212
miasma theory 45
midwifery, Canada 213
migration 93
Milbarth, L.W. on sustainable societies 14
Millard, A.V. on factors affecting child health 182
modernisation, and adaptive behaviour 90
modifiers 83; definition 108
monetary value, problem in our society 74
morbidity factors 40, 81; and socio-cultural environments 105
Morocco: birth control 166–9; reproductive health 157, 158–64
Morrison, B.M. *see* Waxler, N.E. *et al.*
mythology, and scientific research 66

National Academy of Science, USA 46
National Health and Medical Research Council, Australia 59, 74–5
national income, definition 112–13
National Institute for the Environment 66
national resources, and child health 178
National Strategy for Ecologically Sustainable Development 123
natural factors, modifiers 83
natural resources, as environmental services 121
nature conservation: and health 60–63, *see also* conservation
needs, human 116
neighbourhoods: and health 199; and isolation 200
neonatal intensive care unit, survival in 186
Nepal, and traditional medicine 212, 221
Netherlands, temporary food shortages and child health 181

Newcastle, Thousand Families Study
185-7
news coverage 142
Nicaragua, and traditional medicine 212-3
niche, environment of individual 82
non-adaptation 91
North–South divide 6-7
nutrition xvii

Omran, A.R. and epidemiologic transition
theory 155
ontogenetic development 79-80
optimal range 85
organic recycling, Indonesia 227
organisms, and their environment 79,
85-92
organs, ageing of 80
Ornstein, B. and Schatz, H. on
homelessness 202
Ottawa Charter for Health Promotion 42,
44, 48, 49, 50; socio-ecological
approach to health 37, 38
Our Common Future 38, 48-9, 54, *see also*
Brundtland Report; World Commission
on Environment and Development
ovarian ageing 159
over-adaptation 89-91
overpopulation 48, 69
Owen, L.A. 114
ozone layer 37, 54

Pakistan, and traditional medicine 212
paragenetic factor, definition 109
parasites 67
parenting, and poverty 178, 180
Park, Robert E. 7
Park, Robert E. and McKenzie, Roderick,
D. 4-5, 15
partnerships and alliances, effectiveness of
72
pathogenic factors, and phylogenesis
105
pattern of knowledge 47-8
pattern of relations, organisms and
environment 45-6
Pearce, D. *et al.*: on sustainability 113; on
threshold uncertainty 128
pejus 85; definition 109
people, interaction with environment 67-9
pessimum, definition 109
pharmaceutical industry: and reproductive
health 170; and traditional medicine
215, 218
phenotype, definition 109

phenotypic plastic changes *see*
developmental adjustments
philosophical approach 7
phylogenesis 80, 85; and pathogenic
factors 105, *see also* genetic adaption
Pickering, T. and Owen, L.A., on wealth
and health 114
planning for sustainable health 120-27;
and budgets 130; and community
participation 201
Poland: income 95; infant mortality
101-4, 102
polarisation of views, and the media 146
policy making, USA 136
political advertising, and the news 142-3
political ecology 49, 52-7; definition 135;
ideology (USA) 138-42; and impact of
policy on health (USA) 147; political
parties (USA) 146; press (USA)
141-46; public opinion (USA) 136-8,
see also human ecology; political ecology
political ideology 138
pollution 91, 118, 199, 201; and healthy
houses 196; Indonesia 227
polygenic trait, definition 109
poor, and sustainable development 131
population density 68-9, 73; developing
countries 230-32; and environmental
degradation 71; Indonesia 234; and
technological breakthroughs 74
poverty: and child health 178-9, 182;
effect on parenting 180; and
environmental problems 69; and ethical
systems 104; feminisation of 47; and
homelessness 204
pregnancy *see* maternity
press *see* media
preventive measures 155; and child health
177; effectiveness of 71; public interest
in 52
problem solving 14-16, 64; and cultural
values 68
productivity, Indonesian agriculture 227
progress, and gross national product 38
proximal causes of child ill-health 176
psychology 7
public consultation, need for 70
public health: definition 115; ecological
37-8, 45-7; and ecology 52-7;
infrastructure needed 42; and medical
science 39-40; proposed new definition
44; strategy for 39-42, 50-53
public opinion: and information 139; USA
136-8

punishment, and compliance 70

rainforests, Indonesia 234
reafforestation, Indonesia 234
reflexes, unconditioned 86
regulatory adjustments 87; definition 107
Reiter, Hanns 2
relationships, and ecology 3
religious crisis, and 'homelessness' 194
Report by Physicians for Social
    Responsibility 118
reproduction, and health problems 169
reproductive ageing 159
reproductive cancers 170
reproductive health 155–7; behavioural
    determinants of 161–4; biological
    determinants of 157–61; and changes in
    patterns of sexuality 169–71; Morocco
    157, 158–64, *see also* fertility; health
reproductive patterns 162
research, traditional health care 214–18
research methodologies, and traditional
    medicine 216
resources: availability of 71–2; and health
    70
response norm 93
restrictive strand in planning 126, 129
rice 239, 241; cultivation of 233, 242
Rio de Janeiro Conference 6–7, 22, 119
risk: assessment and prevention 70;
    changing social perception of 54
Rodgers, K.P. and Saunier, R.E.: on basic
    health care 119; on integrated planning
    122
Rogers, R.G. and Hackenberg, R. on
    prevention strategies 155

safety, traditional and modern health
    systems compared 216–17
salutogenis 46
sanative communities 74
Sandman, Peter on risks 39
sanitation, and child health 179
Saunier, R.E. 119
scale 63
Schatz, H. 202
schistosomiasis, and traditional medicine
    219
Schnaiberg, A. 46
science as part of the problem 43
scientific medicine 40
scientific research 66; and prevailing values
    62
selectivity, and the media 145–6

self-actualisation 7
self-awareness xvi
self-regulating systems 81
sewage treatment, Indonesia 241–2
sewerage principle, and ecological principle
    43–4
sex hormone levels 159
sexual activity, and age 159
sexuality, patterns of and health 169–71
Simonis, U.E. on basic needs 117
single motherhood 164
Sirisena, W.M. *see* Waxler, N.E. *et al.*
skin cancer, increased rates of 54
Smith, G.D., Blane, D. and Bartley, M. on
    factors affecting child health 182
smoking: and child health 179–80;
    maternal: campaign against 187–8
social development and population
    increase 230
social ecology of water 254–6
social engineering 5
social equilibrium 4–5
social equity 73
social flexibility 48
social investment 49
social medicine 40
social rights, and health 52
social tension, and infant mortality 101
social vs contractual relationships 196
socialisation, and sense of belonging 195
societal developments, and economic
    growth 38–9
Society for Human Ecology (SHE) 6
socio-cultural environments, and
    morbidity factors 105, 106
socio-economic circumstances: and ability
    to parent 178, 180; and adolescent
    offending 185–7; and fertility 165–6;
    and health 185
Solorzano, A. 20
somatic traits, indices of status 96
Sontag, M.S.: on core concepts of human
    ecology 10; on families 18; on
    interdependence 17
Spain: birth control 166–9; reproductive
    health 157, 158–65
specialisation, problems associated with
    11–12
specialist and generalist research 13
spiritual dimension of human ecology 9
Sri Lanka: infant mortality 184; and
    traditional medicine 212
stature, indication of living conditions 95
Stockholm Conference 6

'Strategies for National Sustainable
    Development' 119–20, 123
Strategy for Sustainable Living 112–13,
    125
Straus, D. on integration of disciplines 7;
    on specialist and generalist research 13
stress 48; and adaptational changes 106
stress limits, *see also* Ultimate
    Environmental Thresholds
Sudden Infant Death Syndrome 176
sustainable harvesting of medicinal plants
    222
sustainable health: planning for 120–27;
    and sustainable development 112–20
sustainable society xvi, 14, 23, 62, 63;
    Brundtland Report 53; and health 130;
    holistic approach 12–13; and limits to
    development 127–31; *Our Common
    Future* 49; and planning 120–26;
    policies required 54–5, 71, 73–4,
    119–20; and the poor 131; and
    sustainable health 112–20; and tyranny
    56; and World Bank 123, *see also*
    development
synergistic approach, traditional medicine
    215
systemic measures for health
    improvements 52

Tansley, A.G. 3
technology: and aspirations 72; and
    population increase 74
Third World *see* developing countries
Thomson, J.A. 5
Thousand Families Study, Newcastle
    185–7
threshold to development 127–30
threshold uncertainty 128
tobacco industry 54
Tolba, M. on sustainable development 114
toxic chemicals, and the food chain 54
trade unions 40
traditional health systems 209–10; Africa
    213–14; the Americas 212–13; Asia
    211–12; and biodiversity 218–23; costs
    214–15; priorities 223–4; and research
    214–18, *see also* health
traditional medicines: and AIDS 213–14;
    cultural factors 211; and intellectual
    property rights 218–19, 224; and over-
    exploitation of plants 220–23; and
    safety 216–17, *see also* health
TRAFFIC 220–1
traits, distribution of 99–101

transcendental nature of human ecology 8
tuberculosis, and traditional medicine 214
tyranny, and sustainable society 56

Uganda, and traditional medicine 213–4
UK traders, medicinal plants 221
Ultimate Environmental Thresholds
    (UET) 127–30
under-developed countries 6
urban population, as percentage of total
    population 244
USA: beliefs 137; 'Contract with America'
    136, 138, 147; flaws in environmental
    research 66; homelessness 202; ideology
    and political ecology 138–41;
    information sources 140–2; media
    141–7; political parties 146; public
    opinion 136–8; and traditional medicine
    213

values: cultural 67–8; and human ecology
    13–15
Varea, C. on traditional contraception 169
Victoria Community Health Councils,
    Australia 51

wage labour, cheap female 199
water 250; and culture 251–2, 254;
    Indonesia 235–43; Java 125; quality of
    and parasites 242–3; social ecology of
    254–6; waste and reclaimed 252–4
Waxler, N.E., Morrison, B.M., Sirisena,
    W.M. *et al.* on infant mortality 184
wealth, and health 114
welfare benefits, USA 137–40
welfare improvements, and development
    112
well-being 61, 113–14
Westney, O.E.: core concepts of human
    ecology 10–11, 15; on input-output
    factors 17; self-actualisation 7
wetlands 254–5
Winslow, C.E.A. on public health 38
women: and birth control 166–9; and
    changes in reproductive patterns
    169–71; and fertility 157–66; and
    health 153–5; and reproductive health
    155–7
women's self-perception, and advertising
    170
Woods, R. *see* Blackman *et al.*
work 5
world: and health 21–2; and human beings
    19

World Bank: family planning 232; gross national product 56; strategy for sustainable development 123
World Commission on Environment and Development 38, 49, *see also* Brundtland Report; *Our Common Future*
World Employment Conference on basic needs 117
World Health Assembly 49
World Health Organisation (WHO) 21, 41–2, 193; definition of human ecology 8; on loss of medicinal plants 220; and traditional medicine 209, 210, 214
Worldwatch Institute 56
World Wide Fund for Nature (WWF), concern over medicinal plants 220

Young, Gerald: characteristics of human ecology 8–9; human ecology 2

Zarsky, L. on deterioration of environment 117
zebra mussels 219